Construction Commun

Construction Communication

Stephen Emmitt

Department of Civil Engineering
Technical University of Denmark

Christopher A. Gorse

School of the Built Environment
Leeds Metropolitan University

Blackwell
Publishing

© 2003 by Blackwell Publishing Ltd
Editorial Offices:
9600 Garsington Road, Oxford OX4 2DQ, UK
  Tel: +44 (0)1865 776868
108 Cowley Road, Oxford OX4 1JF, UK
  Tel: +44 (0)1865 791100
Blackwell Publishing Inc., 350 Main Street, Malden,
MA 02148-5018, USA
  Tel: +1 781 388 8250
Iowa State Press, a Blackwell Publishing Company,
2121 State Avenue, Ames, Iowa 50014-8300, USA
  Tel: +1 515 292 0140
Blackwell Munksgaard, 1 Rosenørns Allé, P.O. Box
227, DK-1502 Copenhagen V, Denmark
  Tel: +45 77 33 33 33
Blackwell Publishing Asia Pty Ltd, 550 Swanston
Street, Carlton South, Victoria 3053, Australia
  Tel: +61 (0)3 9347 0300
Blackwell Verlag, Kurfürstendamm 57, 10707 Berlin,
Germany
  Tel: +49 (0)30 32 79 060
Blackwell Publishing, 10 rue Casimir Delavigne,
75006 Paris, France
  Tel: +33 1 53 10 33 10

First published 2003

A catalogue record for this title is available from the
British Library

ISBN 1-4051-0002-8

Library of Congress
Cataloging-in-Publication Data
is available

Set in 9.5/11pt Palatino
by DP Photosetting, Aylesbury, Bucks
Printed and bound in Great Britain by
TJ International, Padstow, Cornwall

For further information on
Blackwell Publishing, visit our website:
www.blackwellpublishing.com

Contents

Preface

Effective communication lies at the heart of business, inherent in leadership and management. Regardless of our individual knowledge and skills, if we are unable to communicate ideas to others, simply and clearly, then we are unlikely to succeed in our endeavours. We are perceived and evaluated by others on the basis of our written, graphical and conversational dexterity. Ultimately we are judged on what we write, draw and say, how we communicate and when we communicate. Communication skills are highly transferable and particularly important in the realisation of a construction project.

During separate careers as architect and construction manager respectively we have been concerned with communication, or rather the lack of effective communication, within the construction process. No matter how thorough the briefing process, how clear the drawings, how good the site management, there were always a number of problems that arose during each and every project, regardless of size or location. Some of the difficulties were minor and easily resolved, some were major and lead to conflict and dispute. With every new project fresh and slightly different problems developed. With the benefit of hindsight and the time to reflect on these problems as researchers, has come a somewhat startling, but simple, observation. In every case the problem could be related back to some form of communication breakdown. By communication breakdown we mean the failure of one party to convey his or her intentions to another, leading to misunderstanding and the associated problems that such a state may bring about.

Within the literature on design and construction management it is common experience to find criticism about the construction industry; how poorly it is structured; how ineffective it is; how poorly people work together; how they fail to communicate and how much better things could be. There is, however, very little practical guidance on how communications in construction can be improved. During our time in construction we have been engaged in trying to improve communications within and between organisations. We have done so by trying to simplify processes (in line with the philosophy of constructability) and through open communication that relies on a degree of trust and mutual sharing of information (in line with the partnering philosophy). Neither approach is easy to achieve in practice when working in a sector renowned for its fragmented and adversarial nature. Indeed, as soon as things start to go wrong it is particularly difficult to keep one's nerve and commitment to open communication, especially when everyone around you is losing their nerve and resorting to defensive communication. We have also been involved in implementing and managing change through process innovations, for which good communication is key. By this we do not just mean the ability to communicate change, but also the ability to understand communication within a work environment before any attempt to implement change is introduced.

For the purposes of this book we have concerned ourselves with investigating communication in the construction process and have attempted to highlight some of the issues with which the majority of readers can identify. Through a greater understanding we hope that communications may become more productive, thus helping to reduce the number and severity of problems brought about by ineffective communication. This may have a direct influence on the quality of service provision, quality of the finished building and (hopefully) help to reduce the potential for conflict during a project's life cycle. There are many comprehensive books that deal with communication and related aspects, although many of the principles do not transfer easily to construction. In this book we have tried to build a bridge between this large body of literature and the social system we know as construction. By combining the industrial experience and research findings of both an architect and a construction manager we have attempted to reduce, if not eliminate, any form of bias towards the design or production side of the building process; essentially the congruence of design and production. In doing so we have attempted to make the text accessible to all those involved in construction, from client right through to building user and facility manager.

We use terms such as building and construction, architect and designer, etc. interchangeably because they are interchangeable in practice, besides which, agonising over precise definitions is usually self-defeating. We have also attempted to follow the advice of Sir Ernest Gower (1954) and to write in plain words. Since we hope to interest those working in the building industry in communication science and those in the field of communication in building, we have dedicated space to some of the more elementary issues of both fields. We hope this book will help to highlight the challenges faced by those working in construction and the underlying issues that colour project relationships. In particular we hope the contents will help to stimulate and inform those charged with managing communications and/or effecting change (regardless of job title or position). Effective communication is the key to success.

SE and CAG
Technical University of Denmark
Leeds Metropolitan University
se@byg.dtu.dk
c.gorse@lmu.ac.uk

1 Construction: a social perspective

The construction industry is not a homogeneous industry, it is made up of many diverse and competing organisations and professional partnerships, the majority of whom are brought together for one, bespoke project, before transferring to the next. The industry is notorious for its adversarial behaviour and litigious orientation and it is questionable as to whether there is ever a real 'team effort' when it comes to designing and producing a building. In this chapter we attempt to explore some of the fundamental challenges inherent in building from a communication perspective, starting with an overview of the development process and the characteristics of the participants.

People businesses

Everyone concerned with the design, erection, use and eventual recycling of a building relies on communication, or more specifically effective communication, to get things done. Initiators and sponsors of building projects must transmit their thoughts and aspirations to the designer; the designer to the construction manager; and the construction manager to tradespeople. This fundamental, yet vital, process we know as communication is frequently taken for granted until something goes wrong. At this point we become more defensive and more aware of the consequences of our communications as attention shifts to who communicated what, to whom and when, i.e. we look for someone to blame. Clearly, it is to the advantage of everyone involved in construction to be able to communicate clearly and efficiently. The greater the empathy between individuals the better the communication and the greater the client satisfaction with the finished building. Conversely, the less effective the communication the greater the likelihood of dissatisfaction and conflict. The consequences of ineffective communications makes for sensational headlines in the trade press and we could be forgiven for thinking that the organisations and individuals involved in construction projects were not particularly good at communicating. This would be misleading. The vast majority of projects progress relatively smoothly, with minor problems resolved as they arise and with project goals being met. Unfortunately, successful projects appear to make less interesting headlines and so our focus is directed towards failure rather than success.

Interpersonal and intergroup communications are vitally important to the success of organisations and individual projects. Unfortunately, over the past decade or so we have become distracted by the lure of information technologies. ITs have developed rapidly and provide a convenient tool through which to transmit, store and access vast quantities of information very quickly. These technological advances are very welcome; however, anecdotal evidence suggests that we have started to focus too much on the power and speed of the systems, and not enough on the messages being transmitted, or the requirements of the users. We have become overloaded with information and spend so much time trying to cope with it that we overlook the importance of human interaction. Indeed, there is a tendency for people to hide behind the perceived safety of their computer screen (relying on email, etc.) instead of conversing face to face, in many cases to the detriment of the project. We need to interact more, not less. Information management and the management of communications is an important area for individuals and businesses alike. So, too, are interpersonal skills for the effective running of organisations and individual projects, because without them it is difficult to get the message across.

Reports urging greater productivity, improved quality, improved service delivery, better value, better safety, greater adherence to programme and so on – for less money and in less time – are a frequent reminder of the challenges we face. Although well intended, the majority of the reports display a surprising ignorance of the complex organisational and personal relationships that make up the exciting culture of construction. Management innovations are not particularly easy to achieve in practice – despite what the management gurus may claim – because they, too, rely on the wonderfully idiosyncratic, individual and unpredictable nature of people. With every new idea or fad comes increased complexity, additional paperwork, more convoluted relationships, even more consultants and even more sub-sub-contractors with the associated transfer of responsibility, i.e. lack of responsibility, and questionable improvements in either service delivery or quality of the finished product. With increased complexity of the communication network structure comes the increased potential for ineffective communication, errors and disputes. Add to this the associated issues of trying to do too much in too little time and the problem is exacerbated. Change is necessary, although before we embark on any programme of change, no matter how grand or insignificant it may appear, we must try to understand the existing relationships within individual organisations and within specific project environments.

People build and these people must communicate with one another effectively in order to achieve their common objective. It is people who commission building projects, who do the designing, schedule programmes, design the project's culture and work together through a variety of communication media towards a common goal, a completed building, be it a small domestic extension or a multi-million pound development. People then interact with the building during its life, altering and adapting the artefact over time to suit changing requirements and trends. Eventually the building is dismantled, materials are recycled and the site is put to another use, i.e. the process starts all over again. We raise this as an issue because construction is not like other industries, and techniques adopted successfully elsewhere need very careful consideration before they are forced on a very different (and often reluctant) sector. Construction is not a homogeneous industry, it is made up of a fascinating mixture of companies and professional consultants, entrepreneurs and tradespeople, all competing to make a living, and usually drawn together for one specific project, never to work together again. This loose coalition of people and organisations will change during the life of the project, so there is never any real 'project team', rather a collection of groups and individuals. The manner in which the project participants communicate with one another, through formal and informal communication channels, is key to a successful project. Communication will help individuals to establish a degree of trust, help to achieve empathy and thus influence the synergy between them. It follows that the faster they are able to communicate effectively the faster they will establish good working relationships and hence the stronger the likelihood of a successful project.

Mechanisation, standardisation and computer technology may have reduced the number of people involved in the process, but we still need people to produce the designs, work the machinery and communicate with one another in order to achieve a common goal, the realisation of a completed building project.

Unfortunately for managers, people are unpredictable, equally prone to moments of inspiration and incompetence, marvellous managers and communicators for 99 per cent of the time, yet hopeless for the other 1 per cent. In construction, with its people-based businesses, the human factor cannot be ignored.

The development process

The procurement of a building is a complex, time- and resource-consuming process. While some people prefer to buy structures that have been built speculatively, such as private house developments, the majority of building projects are designed and built to order, i.e. the product is bespoke, regardless of the amount of prefabrication employed in its construction. Delivery of a well-designed and well-constructed building that is functional and enjoyable to use requires expert managerial skills throughout the entire process, from the brief through to occupation. At the heart of good management lies the ability to create, promote and sustain healthy communication networks. From a communication perspective it is necessary to recognise that a number of diverse individuals and organisations come together for one project, forming communication networks in the process. When the project is complete they will all go their separate ways to join new, often quite different, projects and in doing so will form new networks. Thus, in contrast to some other manufacturing processes that rely on static plant, consistent supply chains and repetition, the relationships in construction are seldom stable and often rather short-lived. It is essentially an industry of organisations brought together for a specific task on a particular site, held together by the project glue – a temporary multi-organisation. The implication, therefore, is that the process of building is very complex and deserves attention if communication is to be effective throughout the entire life of a project, from inception and briefing through to completion and occupancy. So the building, the process and the communication networks that develop for a project are unique.

Characteristics

Clients make increasing demands in terms of improving the performance of their buildings (functionally and aesthetically) while at the same time trying to reduce the initial capital outlay, operational and maintenance costs, and also the time to design and construct the building. Set against an already competitive industry these pressures have tended to result in different ways of trying to achieve objectives. In some respects this has brought about greater specialisation, diversity and, of course, fragmentation, all factors that influence the efficiency of communication between various contributors to construction projects. However, there is also a move towards integrated service providers, the 'one-stop shops' where, in theory at least, individuals are working together and thus the opportunity for efficient communication may be greater than in more fragmented arrangements. In relation to communication there are a number of consistent characteristics.

Project dependent

Because projects vary in size, duration, location and quantity it is difficult to adequately predict workload over the longer term. One direct result of this is the tendency for contractors to rely on casual labour and sub-contractors, a characteristic also present among the consultants, although they tend to use the terms 'contract staff' and 'outsourcing'. Fluctuations in workload brought about by changes in demand, leading to over- or under-capacity, adds further to the lack of consistency. Furthermore, regardless of the amount of prefabrication off site, the materials, plant and labour have to go to the site, i.e. they move from one location to another – both involve logistical issues. Construction project characteristics include:

- Lack of continuity within and between projects, which makes the establishment and promotion of efficient and effective communications particularly challenging. With each project individuals are faced with communicating with unfamiliar organisations and unfamiliar individuals. In such an environment it takes effort and time for effective communication to be achieved.
- Each new project will have different participants, thus relationships and communication channels have to be (re)created for each project. From a managerial perspective, what worked on the previous project may not on the current one. Emphasis is on key individuals to ensure communication routes are in place and are utilised.
- Individual projects are unique in their design and specification, material specifications alter between projects, thus it is difficult to ensure consistent supply chains. This means that new manufacturers may be introduced and hence new communication routes need to be developed for each project.
- Projects can last a long time and during this time participants may change, e.g. moved jobs, and thus interpersonal communication channels will need to be re-established throughout the project duration.

Complex structure

Different organisations are involved in design, engineering, surveying, contracting, plant hire and material production and supply. The sector is multi-organisational (sometimes referred to as multi-party). Each organisation is affiliated to a particular professional organisation or trade association whose concern is to look after their members' interest with little consideration for co-operation and collaboration. The 'project team' appears to be a myth, instead there are a series of poorly connected teams or groups that carry out specific functions for a particular project. Design and construction are obviously separate functions, but so too are many other operations, e.g. design of the structure and design of the services. Furthermore the tendency of main contractors to sub and sub-sub contract work further undermines any real team approach. If the project is well managed it is possible to create a unified approach, if not then the cracks soon appear and organisations quickly resort to an 'us and them' approach, conflict can occur and minor problems become blown out of all proportion, the only winners being the legal profession. Conflict appears to be endemic and difficult to cure. There is:

- no single project 'team' or organisation, rather a temporary (ad hoc) arrangement of different organisations contributing to a particular project at different times as a coalition
- no overall goal, other than shared and/or individual project deadlines.

Temporary supply chains

Construction relies on many different sectors for the supply of services and materials; some of these are dedicated to construction, others supply many different sectors, of which construction may be a small proportion of its business. Design and construction phases can be lengthy, during which time the people involved may change several times. Contact is temporary and co-operation may be difficult in such circumstances.

Essential characteristics

Combined, the factors identified above will influence the manner in which organisations and individuals interact during the course of a particular project. There are, however, more essential characteristics that are fundamental to all design and construction projects.

(1) *The client and the site*   Clients will influence the communication culture within the project framework by setting the budget and the timescale for completion of the works. The type of procurement route chosen will determine formal communication routes and the responsibility of the various organisations contributing to the project. The site will have an influence since its physical location will influence the regulatory bodies (planning and building control) and local participation (neighbours' input, etc.).

(2) *The individual organisations employed to design and assemble the constructed works*   Organisations are rarely stable, their size and organisational culture will change over time. It is quite likely that on projects with a long duration individuals dealing with particular aspects of a project will change jobs, thus affecting the efficiency of the informal communication channels that would have developed. New employees have to acquire a lot of knowledge about the project quickly and establish their own informal communication routes. Although some of these organisations will be linked through formal contracts, others (e.g. town planners) will not be. Organisational communication has tended to focus on aspects of vertical communication, communication travelling up and down the company's hierarchy system. The project requires effective interorganisational communication, in addition to effective organisational communication. Communication across organisations will be affected by contractual arrangements because different procurement routes place slightly different responsibilities on individuals and hence colour how they interact.

(3) *The individuals within the various organisations*   People have to communicate with colleagues and with others in different organisations. For the majority of time this works well, but occasionally clashes in personality occur that can adversely affect a particular communication route. Formal communication routes are complicated by the adoption of informal communication routes (usually adopted to overcome frustrations with formal channels). We can be unpredictable and most of us have a penchant for using informal routes of communication much to our manager's chagrin. It follows that both interdisciplinary and interorganisational communication needs careful consideration. Face-to-face meetings are an important means of exchanging and sharing information through interpersonal communication.

What does this mean from a communications perspective? It means that the study of communication during the construction process needs to take into account the contextual setting of individual projects, namely the characteristics of the people involved, the structure of their organisations, and the relational architecture imposed by the management of the project itself.

The communicators

Sir Harold Emmerson (1962) noted that efficiency in building depends upon the quality of relationships between the client, professionals, contractor and sub-contractors. He also made the observation that cohesion within the building team was lacking. A criticism that could still be levied at the majority of construction projects the world over. Emmerson's observations are important, because building is a people business; thus relationships are critical to the efficiency and quality of the process and the product. Recent reports concerned with improving efficiency in the British construction sector (Latham 1994, Egan 1998, 2002) have put considerable emphasis on integration, teamwork and partnering arrangements. A philosophy based on co-operation and sharing of information for the benefit of both the project participants and the finished building should be applauded. Indeed, it would appear from conversations with contractors and designers that partnering agreements are on the increase, although whether or not they lead to better buildings remains to be seen. There has also been a trend to see construction as a manufacturing process and focus on the supply chain, with parallels drawn from the manufacturing sector and the military; essentially a matter of logistics. Such parallels, while interesting and valid in certain circumstances, can be misleading because they tend to be applicable only to certain situations, these being very large projects or repetitive projects with a relatively stable supply chain. For the majority of us the supply chain is more of a myth than a reality, a collection of disparate links, each with its own special language of communication.

The cast

The traditional cast of characters, reinforced through academic subject specialisation, comprises a client (building sponsor), project manager, designer (architect) consultants, main contractor and sub-contractors. It is common to refer to *the* client or *the* architect, although in practice there will be an individual representing his or her organisation's interests. A good example would be of an architect's office where the individual who takes the client's brief (usually a senior member of the firm) will pass instructions to a design architect who will then pass the design drawings to a technologist who may then pass them to a project architect to oversee the job on site. We shall return to the complexity of communication within organisations later, here we are interested in the relationship between the different parties brought together for an individual project, discussed below.

(1) *Client*   Sometimes described as the building sponsor, this individual or organisation pays for the project. Not surprisingly, clients demand excellent service and high quality buildings at a realistic price. They want value for money. The relationship between the brief-taker and the client is crucial to the development of the project.

(2) *Professional consultants*   Regardless of actual specialism, all professionals rely either directly or indirectly on the client for their fees and hence their livelihood. Their collective task is to take the client's requirements and use their knowledge, experience and skills to convert them into information from which the project can be constructed. Under traditional forms of contract their fees come directly from the client; under other forms, such as design and construct, the consultants' fees come indirectly via the organisation leading the project, i.e. the contractor.

(3) *Main contractor and sub-contractors*   Again, regardless of contractual arrangements these organisations and individuals also rely directly and indirectly

on the client to pay for materials, plant, labour and management of the project.

(4) *Legislative bodies* Building codes and town planning legislation must be complied with. In many situations approvals may be difficult to obtain and communication skills are required to argue one's case and, hopefully, receive the necessary consents and permits without undue delay or loss of design intent.

(5) *Interested parties* Building users, members of the community and special interest groups will want their concerns to be addressed and may influence the design process. Again communication between interested parties and the designers is key to the development and retention of good relationships with neighbours.

Communication channels between parties are dependent upon how the building team is comprised and the procurement route selected. It is also dependent upon the particular stage of the project or operations. The Building Industry Communications publication (1966) highlighted the fact that uncertainty exists between team members, in particular between:

- Client and design team members
- Design team members
- Design team and construction team
- Construction team members.

The publication also raised the problem of uncertainty outside the formally constituted team, from those not directly involved (planners, public bodies, pressure groups, etc.) and from resources (the availability and consistency labour and materials). In all these situations issues of trust, confidence, reassurance, co-operation and diversity of interests have to be considered and, hopefully, accommodated. It is when the contract nears completion that stresses are most likely to be highest as pressures on time, cost and quality mount; the deadline of ensuring practical completion. This very diverse group of people must be managed at different stages in the project's life. In particular, it is where groups and/or individuals interact, the boundary condition, which needs to be effectively managed so that there is no loss in the quality of information transmitted from one group to another. Thus the project manager (whoever it may be) must be aware of group dynamics and responsibilities throughout the project's quite diverse stages.

In Figure 1.1 all parties have the potential to influence the project. The challenge facing the building management team is to interact with the various parties in a positive manner to ensure that the necessary information is produced and used successfully. In reality the situation is more complex than one party interacting with another. Many of the parties will have formal and informal links with others and each will exert either a positive or negative influence on those with whom they communicate (Figure 1.2). However, the task is still the same, managers and designers interact to exert a positive influence on the communication process ensuring action that leads to the completion of a building to the required standards, within time and to budget.

Temporary contracts and contact

The diversity and temporary nature of building brings about its own inherent problems. Organisations work in a changing, and often uncertain, environment.

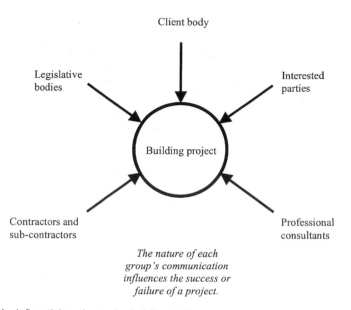

Figure 1.1   Influential parties to the building project.

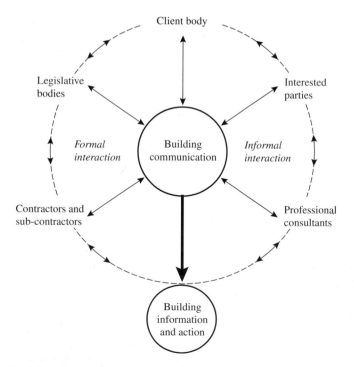

Figure 1.2   Interaction to manage building information.

New materials, technologies and working methods require organisations and individuals to be resourceful and creative – ready to innovate as and when the need arises. Temporary relationships make it very difficult to build good communication networks and they also make it difficult to constantly improve the way we get things done. Consequently it is not easy to constantly improve the quality of the finished product or the consistency of the service delivered. Repetition and consistency have different meanings to manufacturing because contracts, and therefore relationships, are temporary. On small projects the interaction of project participants may be infrequent and short-lived; at the other end of the scale, on large building developments the relationships will last longer. But, as mentioned at the start of this chapter, the communication networks are project-specific and break down once the project is complete. But this statement is too simplistic and potentially misleading because relationships are developing and ending at various stages throughout a project's life cycle.

Procurement systems and professional interaction

No one professional can possess the knowledge required to design and construct a building, thus the temporary project team must seek to pool all of the relevant (and available) expertise at a given point in time to realise the building. Contracts have been developed in an attempt to ensure the parties are aware of, and legally committed to, their roles and responsibilities. A wide variety of publications deal with the procurement of buildings and the many different methods available, so we do not intend to dwell too much on this area. However, it is important to recognise that the type of procurement system used will influence the manner in which the design and construction phases are organised, and hence how individuals interact and communicate through various communication channels. The type of system used will dictate the responsibilities of the client, designer and contractor and their level of control over the process. In some respects the choice of procurement route is about control and power over the project, information, communication routes and decision-making. Basically there are four options, those that are led by either clients, designers, contractors or managers.

(1) *Client-led relationships*   Common on very small projects (e.g. house extensions) and self-build projects. The client may employ a designer to achieve planning consent and building control approval before employing a contractor to construct the design. All communications are via the client.

(2) *Design-led relationships*   Usually referred to as the traditional system of procurement, the design-led form of procurement has evolved over centuries. The client needs someone who can express his or her desires in a design and the architect was the first point of contact. From this the architect provided a professional service, designing the building, appointing the contractor and administering the contract. Historically, traditional contracts were completed sequentially with each stage of the design process being completed before the next commenced and with design work completed before the construction phase; however, it has become common for the process to be accelerated by overlapping the stages, known as fast-tracking. Formal communication routes are determined and controlled by the designer.

(3) *Construction-led relationships*   Design and construct (design and build) is a contractual arrangement whereby the contractor designs and builds a project for a sum inclusive of the design fee and construction costs. Few contracting organisations possess their own design section; instead the contractor employs

design consultants on a fee basis, to undertake the design elements of the package. This one-stop service has gained popularity over the past two decades. Formal communication routes are determined and controlled by the contractor.

(4) *Management-led relationships* Management contracting, construction management and management services have gained popularity over recent years. The management contracting procurement method varies depending on who offers the service, be it a professional practice or a contractor. Such systems normally involve the management of the whole project, from design and production through to the management of the building after occupation. A management fee is charged, normally based on a percentage of the project value. Management systems may provide contract for service (the management of contracts for the client organisation), or the supply of a building (a direct contract to produce a building). Formal communications are determined and controlled by the management organisation.

Each procurement route offers various advantages and disadvantages, the scale of each dependent upon individual circumstances. More importantly, whatever the method of procurement chosen, it is likely that the same professional groups will be involved; it is their contractual and organisational relationships that will differ, as will the communication networks that develop. Separation of design from production has been highlighted as a problem in achieving quality because of the communication barriers that exist (and may still exist in the procurement routes reviewed above). One way of overcoming the potential problem of conflict and ineffective communication is to build using a limited number of intermediaries. An architect-led method of reducing the number of intermediaries is to use construction management, a procurement route that allows the architect to communicate directly with trade contractors and eliminate the main contractor (see Emmitt 1999). However, trying to improve communication through the reduction in the number of (competing) intermediaries is difficult to achieve since the designer must be in a position to influence the procurement route. The argument for effective communication within an information-driven environment is a powerful one, but once again it comes back to the issue of control.

Recognising communication breakdown

The process of communication and timely transfer of information is the key to effective co-ordination and control of the project. Information is required to enable the planned processes and to control change when reacting to the unpredictable elements of construction. Information management is a complex issue, and an area that is starting to receive more attention with the rapid development of IT systems. As the construction team are usually only together for one project, interpersonal communication is required to support industrial relations and develop effective working relationships. Research has found that when communication between team members is most needed, during times of uncertainty and crisis, it often breaks down. The challenge for all those involved in construction projects is to recognise the signs of communication breakdown and try and act before it becomes a problem. Communication breakdown can occur in a variety of guises, from relatively minor instances to more major (and more noticeable) events. In some respects the breakdown in communication can be attributed to the characteristics we noted above; however, many failures to communicate effectively are common across all industries.

Communication matters

Communication is essential to all business activities; it enables organisation, and is an integral part of the construction process. There are very few management development programmes that do not include effective communication as a key skill for effective management, the argument being that any improvement in communication can improve an organisation's operating effectiveness. 'Good communication within an organisation and between organisations contributing to the construction project can improve motivation levels and improve the production process. Conversely, inadequate communication can result in a demotivated workforce and lead to problems in production. Construction projects are complex and risky, requiring the active participation of all contributors. Co-operation and co-ordination of activities through interpersonal and group communication are essential in ensuring the project is completed successfully. Poor communication, lack of consultation and inadequate feedback are to be found as the root cause of defects in many constructed works. Poor co-ordination and communication of design information leads to design problems that cause design errors. Communication is *the* one aspect of the management of projects that pervades all others.

In the chapters that follow we have confined ourselves to an investigation of communication between project participants which naturally leads to a focus on project management. It is important, however, not to lose sight of the aims of the project – to deliver a functional building that people enjoy using. The project is merely a means to an end and we should not overlook the importance of the finished artefact, the constructed works. However, we must recognise that the multitude of decisions made during the contract will affect the finished building and much of the information connected to the building will live on long after the project has been forgotten.

2 Communication in construction

Communication is implicit in everything we do. We all recognise the importance of communication as a tool to achieve our objectives, yet research into communication in construction is scarce. In this chapter we provide a brief overview of the literature including a re-examination of Higgin and Jessop's much cited pilot study. We then explore more recent work, which leads into an overview of construction management research. The chapter concludes with an overview of the various frameworks for administering design and construction projects.

An overview

As noted earlier, the construction industry has a poor reputation for the manner in which its organisations and individuals communicate with one another. Successive governmental reports (Emmerson 1962, Banwell 1964, Latham 1994, Egan 1998, 2002) have consistently drawn our attention to the apparent lack of effective communication within the construction sector. These reports have also highlighted the fragmented nature of the sector, lack of co-ordination, separation of design and construction activities, lack of trust, and adversarial relationships; factors that hinder rather than promote effective communication. These sentiments are echoed in publications produced by the Tavistock Institute in the 1960s (Higgin & Jessop 1965, Building Industry Communications 1966). At the root of these reports is a desire for a more efficient and hence more profitable sector, a concern shared by governments and contributors to construction throughout the world. In essence the reports are a call for greater co-operation, integration and teamwork.

Early work

In the reports mentioned above, there is both an explicit and an implicit charge that poor communication has been a core problem for many years. This is a point which has been picked up by many textbook authors, although paradoxically it is an area in which guidance and advice are lacking. It is clear that the way in which construction activities are organised has a direct affect on communication effectiveness, and vice versa. In an overview of government reports, Wild (2001, 2002a, b) claims that the first specific review of construction communication was undertaken by Higgin and Jessop (1965) and continued by Building Industry Communications (1966). Others (e.g. Nicholson 1997, Smith & Wyatt 1998) suggest that these investigation tentatively follow on from some issues raised in the post-war reports. Wild's review provides a good starting point for highlighting the issues raised, trends and differences between the reports.

The Simon reports (1944, 1945, 1948) dealt specifically with the distribution of building materials, awareness of the problems faced and the contingency planning required to overcome post-war problems. Phillips (1950) picked up on the issues of planning and identified the main improvements required to better understand and manage construction effectively. Issues raised included co-ordination, organisation, planning, the use of specialists and sub-contracting trades, mechanisation, standardisation, financing, contracting, and the role of professionals and clients. The Emmerson report (1962) also highlighted problems associated with co-ordination, planning and management; however, the report implied that some aspects of the earlier reports had been implemented, praising the industry for its flexibility and the operational and technical advances that had been achieved. When dealing with inefficiencies, Emmerson suggested that the industry was fragmented and there was a lack of integration between design and construction. The attention given by Emmerson to the interaction between professionals and organisations resulted in a greater focus on professional and organisational relationships. Although the Banwell report (1964) on contracts and communication continued this focus on issues of formal relationships, Higgin and Jessop (1965) and Building Industry Communications (1966) clearly sought to identify the type of interaction, relationships and groups that make up construction's social systems.

The reports by Simon, Emmerson and Banwell all attempted to encourage collaboration and hence improve information exchange. The Emmerson report (1962) identified the need for co-operation and cohesion and the need for improved communication between parties to the building process. In many respects these observations were echoed in the RIBA's report *The Architect and His Office* (1962), which was concerned with improving the efficiency of architectural practices through better management. This theme was picked up in the first book to address architectural management, by Brunton *et al.* Published in 1964 the book aimed to provide advice to architects to help them with the management of individual projects and their offices, and in which communication was seen as the most important factor.

Higgin and Jessop revisited

Perhaps the most widely cited study of communication in the building industry is *Communications in the Building Industry: The Report of a Pilot Study* (Higgin & Jessop 1965). This report, along with the less well cited *Interdependence and Uncertainty* (Building Industry Communications 1966), came out of the Tavistock Institute in the 1960s, a period during which efficiency and profitability were high on the political agenda. The Tavistock publications helped to highlight the increase in fragmentation and the fact that each specialism had developed its own 'language'. We were left to ponder how anyone managed to build given the difficulties identified. But continue to build we did.

Communications in the building industry

Gurth Higgin (a psychologist) and Neil Jessop (a statistician) sought to investigate the 'dissatisfaction' with communication between members of the building team. They recognised that because of the complexity of construction communications it was difficult to state clear research objectives, and hence undertake long-term research. Hence a three-month-long pilot study, comprising a literature research and an elementary postal questionnaire, was undertaken.

They suggested that the nature of relationships was the main factor behind poor communications, a result of the historical development and fragmentation of trades, professions and responsibilities. This had led to strained relationships, tension and defensiveness when entering new relationships. Second, they suggested that any attempt at improvement was unlikely to yield any degree of success without more information (for which further research was required). The Simon and Emmerson reports were criticised because neither report had led to anything other than minor improvements. Higgin and Jessop discussed the challenges of operational research and the use of critical path techniques before coming up with two hypotheses, which they confess are based on common sense and which are still relevant today (they did not test them). They were:

(1) Co-ordination of both design and construction is better when carried out by a single person (or organisation) than it is when the functions have separate co-ordinators. An early argument for single point responsibility.
(2) If design and co-ordination do have separate co-ordinators, then it is best to ensure early exchange of information. This is essentially a plea for better communication between the designers and the constructors.

Part two of their publication listed five main problems with construction. They were:

(1) Communication with prospective clients could be improved through better client targeting and communicating the range of services on offer to them.
(2) Communication between clients and consultants needed to be improved. Essentially a call for better client briefing before design commenced.
(3) Communication within the design team needed some attention. There was not enough intercommunication between team members, thus objectives were not shared and the process was less effective than might otherwise be the case. Essentially a call for better design management.
(4) Contractual information was deemed to be inadequate, leading to communication difficulties.
(5) Communication within the construction team was seen to suffer because insufficient information was available. Information was incomplete, rushed and not available in time.

From this position they went on to claim that construction was a complex operation and one lacking in information about how construction proceeds (some would argue here that they, as outsiders, did not fully understand the process). They concluded that construction is a series of interdependent operations (again a commonsense observation).

It may be useful to comment, briefly, on their postal questionnaire. Their questionnaire asked respondents to rate the social status of building team members and their contribution to the building process. From 97 responses they found that architects had the highest status, but builders/contractors were perceived as making the biggest contribution – hardly surprising results given the culture of the day.

The fact that this report is cited so frequently may be that little else was available at the time. Earlier work by Bowley (1966) raised more important concerns about structure and communications. Higgin and Jessop ignored her work, although their call for improved organisational effectiveness in the epilogue did echo some of Bowley's earlier observations.

Interdependence and uncertainty

Higgin and Jessop's pilot study led to a more detailed publication by Building Industry Communications (Higgin and Jessop were members of the committee) *Interdependence and Uncertainty: A Study of the Building Industry* (1966), which made a valiant attempt to unravel the complexity of relationships within the industry. The report looked at the industry's structure and put forward practical suggestions for improvements based on the findings of interviews and 13 case studies. The extracts from their case studies clearly illustrate problems related to poor programming, control and communications. In particular, contractors reported difficulties in reading drawings produced by architects and engineers, primarily because they were incomplete. Poor co-ordination of information was another area of concern.

The report described an industry in which abortive work, misunderstanding and delays resulted from failures in communications and division of responsibility. Conflict, confusion, doubt and error figure highly in their picture. Decisions were found to have a knock-on effect down the supply chain, illustrating interdependence within the process. Uncertainty and interdependence were seen as key characteristics of communication and information flow. Interdependence was interpreted as the relevance of different streams of information to each other. Segregating each task was seen as wasteful because new information had to be generated at each interface. This was complicated further by the fact that each participant brought their own experiences and prejudices to bear on the problem, with decisions being taken on an individual level, rather than an industry-wide level.

They also recognised that organisations are not static, and when communication flow is blocked, different organisational groupings develop compared with those when communication is integrated (flowing), the overall recommendation being that construction required collaborative leadership. Problems of control and communications existed within all contractual arrangements, whatever their structure and regardless of who was actually controlling them, be it professional consultant or contractor. The report also highlighted the need for an appropriate scale to measure whether one form of procurement was better than others. However, they also noted that no single characteristic, e.g. as time, cost, or function, would be sufficient to ensure an adequate comparison could be made (they failed to recognise that each project is different from that which preceded it, thus comparisons are rarely robust). The publication concluded that making communications serve the customer would benefit the industry. Seven measures were proposed to ensure ideal communications, namely:

(1) Careful assembly of a multi-skilled team with managerial, technological and analytical abilities;
(2) Removal (or reduction of) artificial barriers, thus designers become part of the site management team;
(3) Considered use of management tools to ensure programming and progress data is continually revised and available to all parties (a task now facilitated by ITs);
(4) Abolition of conflicting interests, through incentives to minimise defensive action (there was no indication as to what the incentives may be);
(5) Adequate resources for obtaining information held off site (again, now facilitated by ITs);
(6) Limit disruption brought about by other projects, i.e. work on one project at a time;
(7) Record all events and actions for later analysis and feedback into future projects. This task should be separate from keeping managerial records. (Essentially an early call for a quality management system.)

A catalyst?

A natural assumption would be that the Tavistock publications formed the catalyst for further research and led to improvements. We are unable to find much evidence of the former and can only conclude that in spite of all the reports and rhetoric there has been little improvement. While the recommendations of the Tavistock reports are still relevant, from a communication perspective the situation has become far more complex. Since the 1960s the number of specialists has increased, the construction industry has fragmented further, technologies have become more complex, the industry has become more litigious and the use of specialist languages (codification) has increased significantly, as has the amount of information required to construct a building. Time pressures are even more of a determining factor than they were in the 1960s due to commercial pressures and the constant pressure on fees and profit margins. On a more positive note, the implementation of quality management systems and the rapid developments in IT systems has helped with the transfer, storage and tracking of resources, information and decisions.

Later work

A couple of publications from the 1970s were also influential. Geoffrey Broadbent's *Design in Architecture* (1973) dedicated an entire chapter to the issue of communication and was important in raising awareness of communication to several generations of architects. John Paterson's *Information Methods: For Design and Construction* (1977) provided a very convincing argument for a simpler way of building through effective management of information. Patterson suggested that it would be prudent to reconsider the problem posed by building, rather than 'sophisticate' our current solutions. He went on to claim that it was essential to look at the whole, rather than the parts, if communications and information flow were to be improved. The reasoning here is that by improving one part (or process), the interface between the remaining parts (or processes) can be rendered more inefficient because of lack of compatibility. Indeed it can (and does) lead to increased specialisation, hybrid forms, fragmentation and, in many cases, reduced efficacy. Construction used to be, and still could be, a simple process. It has become more complex through the increased reliance on information (and hence quantity of information), reduced number of tradespeople (who needed less information) and increased specialisation (increased bureaucracy, legislation and management). Combine this with the desire for control of the process (hence establishing, maintaining and increasing market share) and an ever-litigious environment (another information generator) and we have a communication culture that is more demanding than it might otherwise be.

There has been some interest in communication research within the International Council for Research and Innovation in Building and Construction (CIB). In particular, Working Group 96 'Architectural Management' has witnessed several attempts at dealing with communication, both explicitly and implicitly (e.g. Nicholson 1992, Emmitt 1999) and Working Group 102 'Information and Knowledge Management in Building' has carried out some important work in this area. Additionally, the papers and debates published in the proceedings of the Association of Researchers in Construction Management (ARCOM) conferences, which focus on communication have steadily increased, (e.g. Dainty & Moore 2000, Gorse *et al.* 2000a, b, c, d, 2001, 2002, Hugill 2000, Wild 2001a, Green *et al.* 2002, Moore & Dainty 2002).

Construction management research

Construction management research can be traced back to the early 1960s with initial studies focusing on hard issues. It was not until the mid-1990s that the softer social sciences came to be applied to the construction management field. During the evolution of this research discipline there have been very few publications that have dealt with communications. On the face of it this may be a little surprising. Construction relies on professional interaction and communication across organisational boundaries to develop and implement construction projects. There is no escape from the fact. Yet to observe, model and analyse communication behaviour in construction projects is particularly difficult. This is confirmed in the work of a small number of researchers who have attempted to study aspects of communication behaviour in construction through doctoral research (e.g. Wallace 1987, Gameson 1992, Pietroforte 1992, Bowen 1993, Loosemore 1996, Emmitt 1997, Hugill 2001, Gorse 2002). All of this work was concerned with interpersonal and inter-organisational communication, each doctoral study dealing with a different aspect of communication, for example Loosemore investigated communications during a crisis on construction sites while Gorse researched communication between designer and contractor during site progress meetings. Their work is discussed further in Chapter 14.

A few things emerge from the studies of communication during the construction process. There is a lack of research that observes interaction of the construction manager and other key professionals. The studies are predominantly design orientated with only three of the studies addressing issues that occur during the construction phases. Only Pietroforte's (1992) research addresses the relationship between communication behaviour and the contractual conditions in the USA. Staying with the USA, a study commissioned by the Construction Industry Institute into the effectiveness of communications within project teams concluded that the major obstacle to project success was the 'lack of effective communications' (Thomas *et al.* 1998). Their research was based on analysis of 582 questionnaires completed by individuals representing 72, mainly large, projects. Their conclusions were consistent with the earlier work reported above.

Emerging themes

Hill (1995) found that the diversity and complexity of communication processes does not readily fit with any recognised organisational models. Fragmentation appears to work against the adoption of more effective organisational structures. The consistent theme is the call for improvements to communications. This is relatively well documented. Exactly how this should be achieved is harder to find. Suggestions tend to revolve around some key and, we would argue, misleading areas.

First, is the issue of contractual arrangements. New forms of contract and new procurement routes have been introduced that attempt to remove some of the organisational communications and so promote better teamwork. In the majority of cases the response has been to adopt the new contract but to continue to work within established strictures. Although a few exceptions do exist, we find a situation that is more complex than it was 50 years ago, yet underneath the complexity lies the same fundamental organisational relationships and potential barriers to communication. Second, is the rather optimistic view that information technology will transform the way we work. It will, and has started, to alter it, but transformation takes more than the implementation of hardware, it also requires more attention to the softer, people issues. Finally, it is worth noting that change will come about only if it is in everyone's interest to change. Clearly it is beneficial to some organisations to work in a sector that is fraught with problems. We are talking about contractors that make their profits through claims for additional work and of course those in the legal system who stand to profit most from the mistakes of others.

It is clear that communications are fundamental to the construction process. Although the pilot study by Higgin and Jessop's (1965) and the study that followed (Building Industry Communications 1966) identified communication problems in the construction industry, research into communication within the industry is scarce. The process of communication is about the transfer of information to inform parties and influence action. While government (Latham 1994, Egan 1998, 2002) advocates the need for increased performance and teamwork, we cannot address these issues without considering the nature of communication and its effects. The following chapters gather together the findings from communication research, helping to inform those who are tasked with improving the construction processes. Clearly construction communication research is in its infancy, and we must seek to learn from those social scientists and industrialists, from other sectors, who have recognised the importance of communication for some time.

Frameworks for construction communications

Before proceeding further we need to have some form of framework or reference point. The development process starts with the realisation and identification of a need by the project sponsor, the client. This 'need' may be for a new building, the extension of an existing building, or the alteration of an existing structure; invariably it leads the client to approach and choose appropriate consultants to assist with the design and construction operations. In order to understand the communication process a framework may be helpful.

In the 1960s the RIBA published the plan of work which identified a series of sequential stages through which the project progressed, from inception through to completion. The plan of work was widely adopted by the construction industry and continues to be used as a familiar framework despite some more recent revisions to it. The plan of work is useful for identifying the main stages in the construction process and is applicable to projects that follow traditional procurement routes with the architect appointed by the client and the contractor selected by a competitive tender system. Many projects operate under fast-track methods of design and construction where it is common practice to commence the building works before the design is completed, with elements of design and construction running concurrently. Other models have been proposed, for example Higgin and Jessop's (1965) more simplistic eight phases of building model:

Phase 0	Client deciding to build
Phase 1	Client consulting with team members
Phase 2	Investigating and preparing the brief
Phase 3	Sketch plans, obtaining outline approvals
Phase 4	Preparing contract documentation and obtaining final approval
Phase 5	Agreeing contract and setting up construction team
Phase 6	Construction to completion
Phase 7	Handing over and settling final account

During these phases formal and informal communication will take place between individuals and organisations who are party to the contract. It is the transfer of design intent, the communication of abstract ideas, into the physical building that is a prime concern, a point taken up in Chapter 3.

3 Communicating abstract ideas

At the heart of a successful project lies the ability to communicate abstract ideas from the design office to the site *and* the ability of those on site to translate information into a physical artefact. In this chapter we start to explore the complex languages that exist within the construction industry and the challenge of achieving accurate communication of information. Knowledge assets, abstraction and the codification of knowledge are considered before turning our attention to the challenge of communicating across boundaries and the need for an appropriate language. The chapter concludes with a brief look at the influence of time on the communication process and the physical environment in which it takes place.

Communication and information

Communication and information management is a prime activity in construction. The entire construction process relies on vast quantities of information being generated, transmitted and interpreted to enable a project to be built, maintained, reused and eventually recycled. More specifically, construction industry participants are concerned with information exchange, dealing with drawings, specifications, cost data, programmes, plus other design and management information required for the successful completion of a building. Successful knowledge-based organisations have been shown to rely on the effective transfer of information (e.g. Winch & Schneider 1993, Boisot 1998), and similarly good relations within a team or group are dependent upon effective communication. Problems have been identified in relation to the ease and effectiveness of communications even in small 'communication circles' where the process is relatively simple and the opportunity for interference is relatively low. In construction the information is usually prepared by individuals from diverse backgrounds, such as architects, engineers, subcontractors and specialist suppliers, often using different terms and methods of graphical representation. Thus, verbal communication between two or more individuals is often concerned with resolving queries over the interpretation of the information provided.

Professional interaction and communication

Although general management texts identify communication structures related to traditional and design and build type contracts, research has shown that communication during the construction process does not always follow the theoretical structures proposed in the various guides to managing projects. Pietroforte (1992) found that interaction between professionals was different to that set down in contracts (and assumed in textbooks). A critical comparison was made between the assumed relationships and those observed during the research period. Much of the process was based on informal relationships and casual roles with the exchange of small amounts of information between participants to aid understanding. Hill

(1995) found that formal communication routes were ineffective resulting in the use of informal channels, primarily to reduce the time required to get information, and hence allow the work to continue without delay.

Decisions and interaction

Decision-making is an essential part of both the design phase and the construction process. Interaction between participants during the construction process is necessary to make well-informed decisions. The nature of the interaction and the decisions taken during the project will ultimately determine the success of the process and the quality of the finished product. Communication and the availability of accurate and current information are central to the decision-making process. For the designer, the emphasis is on using knowledge and information to generate creative ideas, from which decisions can be made. For the contractor, the emphasis is on problem-solving through the systematic reduction of the available options. Research has indicated that most construction-related decisions, typically those relating to design and estimating, were not comprehensively informed, due to an over-reliance on self-informing strategies by the decision-makers (Mackinder & Marvin 1982).

Different types of decision-making process can be classified by the information that flows through them (Loosemore 1992). Problems have vertical and horizontal components. The procedures and information created at the various levels in the organisation will influence the decision-making within it. Each type of decision is associated with a level within the hierarchy of the temporary organisation, as illustrated in Figure 3.1. Policy decisions are the first type of decision process and are the most important because they decide the nature of the organisation, the purpose and what it aims to achieve. These are the highest-order decisions and are taken by senior company representatives, e.g. board members and directors. Once the policy decisions have been made, a strategy will be developed and implemented in accordance with the company policy. As part of a client organisation's strategy to

Figure 3.1   Decision process, systems theory, and hierarchy of the decision process.

expand its business new buildings may be required. It is the strategic decisions that determine the type of building, contractual arrangements, design programme and build time and which also initiate the construction process. Deciding on resource allocation attains the tactical goals, which are found on the next level. Tactical decisions will set down the process that will take the project from inception to completion The resources allocated in the tactical process must be co-ordinated to achieve the goals that have been established at the strategic level. The co-ordination of resources is processed at the operational level of the decision-making process. Lower-level decisions must, however, operate within the constraints of the higher ones.

Under traditional contracts the construction manager would not be appointed until the strategic decision-making process had taken place. Most of the decisions made between the architect and construction manager will be tactical and operational. The systems theory has been expanded to include subsystems. This can be used to identify and analyse the interrelationships of the component parts of the temporary construction organisation. Subsystems, such as the main contractor, enter into the construction project as a result of the strategic decisions made by the client organisation; the highest level decision that the subsystem can make is a tactical one. Tactical decisions are very important because they will influence the quality of the final product. Tactical and operational decisions can also affect completion times and costs, the decisions made by construction managers and designers could impact on the success of the strategic and policy decisions made by the client organisation. The majority of the decisions made during the construction phase and management and design meetings will be tactical decisions.

Ineffective communication and conflict

Ineffective communication has been identified as a problem that can lead to conflict and subsequent litigation. Analysis of legal cases has shown that building failures can be traced back to a mismatch of knowledge and expectations (Lavers 1992). Poor communication may result in a quality of service delivery being below the specified standard and may also result in buildings that fail to meet the specified performance requirements. Specialists are employed because they have the knowledge and experience required to complete a specific set of tasks. The production of a building requires the combination of the knowledge, skill and experience of many different professionals. Professionals cannot know everything, and there are times when specialists will not have an adequate understanding of certain components or procedures. Thus they need to seek information from others and ask questions in order to reduce their knowledge deficit. Unfortunately, question-asking behaviour can be perceived to denote a lack of competence and some professionals are reluctant to ask for advice or admit that they do not understand. Failure to ask questions and/or admit that more knowledge is required will lead to problems. As a consequence it is not uncommon for a mismatch in understanding to occur, and it is the role of interpersonal communication to reduce the disparity in knowledge.

Clarification constitutes good practice and is part of the legal obligation of construction professionals (Lavers 1992). There is a need to ensure that when decisions are made, the agreed objectives are understood by all and their effect on the final product is anticipated and understood by all project participants. Expectations and outcomes of the decision-making process, which are manifest in each stakeholder, should be clear and relevant, thus helping to avoid conflict. Lavers suggested that greater attention should be paid to improving communication as a means of eliminating disputes and potential claims.

Information flow and knowledge management

Drawings are the main medium used to transmit the designer's intent to the contractor; however, the format and intent of the drawing is often far more apparent to the originator than it is to the receiver. It is not uncommon for the receiver to request clarification or even misread the originator's intentions, sometimes with costly consequences. The effect is magnified when several drawings from different originators are being referred to at the same time. It is rare, even with the use of digitally generated drawing systems, for architects, structural engineers, electrical and mechanical engineers to use the same symbols and terminology, thus co-ordination is a constant challenge for the user and especially the co-ordinator. The generation of drawings within the originating office is a process that relies on the use of information and knowledge, much of which will not be included in the finished drawing. Such information must be managed within the office and the quality of the resultant documents checked and controlled before it is communicated to other parties. Construction is about information transfer, exchange and use, and it follows that information flow will be a primary concern of the project co-ordinator or project manager, as will the transfer and control of knowledge.

Information technology can help to improve communication but its development must take into account the social complexity in which information is exchanged. The problem is not so much with the speed of delivery, but more in the quality of the information delivered and the managerial structure of the communication networks. IT systems must be managed in the same way as well-designed paper-based systems are, i.e. they must be managed to ensure the information circuit has value to the users. From inception to use, reuse and disposal, people are involved, and the manner in which they communicate (or fail to communicate) has a considerable impact; as such, it is necessary to look at the whole process, rather than the parts, if communication and information processing are to be improved (Paterson 1977). Communication and information are inextricably linked and need to be addressed as integral, not separate issues. Human beings are data-handling machines, each of us varying in efficacy, each of us different because of our different life experiences. Heuristically, we are good (in theory at least) at handling information; however, most of us are poor at holding a lot of information in our memories. Computers can now do the memory task for us very well; the heuristic tasks are still dependent on people. Our problem comes in knowing what to access, what to ignore and what to transmit. Writing about knowledge and information in architecture, Paterson (1977) makes two observations:

- knowledge is infinite (we can never get to the absolute truth); and
- data is environment dependent, i.e. it will have different meanings when placed or taken from different environments.

Paterson (1977) provides a simple model based on information flow. The five functions are:

(1) Information (input, storage and output)
(2) Design (analysis, synthesis and evaluation)
(3) Communication (constructional and financial)
(4) Construction (construction management and financial management)
(5) Maintenance and control (access to 'as built' information)

Throughout the project's life cycle, information will be generated, stored, discarded and transmitted through a variety of communication media and communication channels. In doing so there will be a continual input of new knowledge (and of course the loss of existing or old knowledge). Experience gained from working on projects should go some way to enhance not only the knowledge of individuals but also the collective knowledge of the organisation. New projects will start from a different base line to that which proceeded it.

Definitions

We have a symbiotic relationship between information and knowledge, both of which require sympathetic management to enable the realisation of organisational and individual effectiveness. There is also a very clear relationship between information, knowledge and communication. However, before proceeding further it is useful to make a few definitions to avoid confusion. For the purposes of this book the following definitions are used.

Information and data

For practical purposes information and data are artefacts that add to our sum of knowledge on a particular issue. However, philosophically, these are environmental occurrences that have the potential to become sensory signals. Due to our knowledge and cognitive ability the information and data may mean different things to different people. However, through common experiences, education and training, the human intelligence enables us to recognise and communicate information that has congruence (has almost a common meaning). Thus, most construction professionals will have a good and almost common understanding of the design information used. As construction information becomes more specialised the ability to achieve a common understanding among all parties becomes more difficult and requires more effort. For example, an engineer's calculations may prove that a steel frame can transfer building loads safely to the foundations, but the architect and contractor will not be able to check or understand the calculation unless they also have the specialised knowledge (which is unlikely, and arguably unnecessary). This means that professionals must rely on others party to the project and trust them with their particular level of expertise.

Communication

Any act or event that a person perceives can be deemed to be an act of communication. It may be information gained from verbal and non-verbal information, body language, facial expression, touch and olfactory information from our immediate environment that is made manifest and therefore has meaning. Some schools of thought take the view that when we enter a room or environment where nobody else is present we are merely processing environmental information, thinking and feeling but not really communicating. For most scholars the act of communication only occurs between two or more people. Communicators use utterances, signals and contextual clues which have relevance to the situations. An effective communicator intends to produce a relevant utterance or signal that creates a contextual effect (induces understanding and reaction) that requires minimal processing effort by the receiver. Communication starts with an act or an event. The communication act or event provides information that has relevance and meaning to the person or persons perceiving it. The information produced may not have the same meaning to different people and may not result in the same outcomes (manifested in behaviour, action, influence, etc.).

Cognition

The process of transforming (reducing, elaborating) and contextualising (assimilating) sensory information to enable understanding, storage, recovery and use is known as cognition. Cognition is not achieved when sensory information stands on its own, is not relevant to previous information and cannot be contextualised. When new sensory experiences are received they must be assembled with other related sensory information. If a professional uses a term or refers to something that another professional has no previous experience of then it will not be understood unless the person receiving the message is able to recognise links within other information provided to them combined with their knowledge.

Knowledge and knowledge management

Something known that has more relevance and contextual meaning than something manifest, assumed or merely experienced is referred to as knowledge. Knowledge is a group of information, facts and framework of thoughts that are objective or verifiable. Through cognition, information is assembled and stored, and when a related topic emerges in a new environment the relevant information can be recalled in the form of knowledge. Knowledge is often classified in two ways: the tacit knowledge of individuals, which is implicit and unarticulated, and the explicit knowledge that is codified and easy to transmit. The importance of knowledge management within organisations is growing. Knowledge management is the process by which information is created, captured, stored, shared, transferred, implemented, exploited and measured to meet the needs of an organisation (e.g. Boisot 1998, Egbu *et al.* 2001).

Information flow

Good communication and information flow is essential if a client's requirements are to be translated into a competent design and well-built product. For example, architectural firms need easy and rapid access to a wide range of up-to-date information. Architects' informational needs will vary through the different stages of the project and must be carefully controlled to ensure that information is both up-to-date and relevant. Carefully implemented filters and controls are required to avoid information overload. Speed of access to relevant information is vital to both the efficient management of individual projects and the efficient use and maintenance of the building and its services.

Information transfer

IT developments have led to the ability to communicate from geographically remote locations. Internet technologies provide an effective tool to manage and disseminate large quantities of information, via intranet and extranet, and to communicate with one another, via video-conferencing. Intranet is a closed communication network that allows individual users access to all information on the system. It is commonly used by organisations to assist their employees in their job, with a very limited amount of access to external organisations, e.g. regular suppliers. Intranets have been used for managing project information, i.e. they are set up for the life of a project, thus allowing participants access to the project information. At the end of the project the project intranet is shut down, often with the loss of the collective knowledge gained through the experience of delivering that particular project. Project extranets are more sophisticated. Extranets are hosted by a particular organisation, e.g. project managers, that are already using intranet technology. Organisations contributing to a particular project are allowed access to a certain part of the host organisation's intranet in order to share information and collaborate more effectively. This system also allows project partners access to selected parts of the host organisation's collective knowledge base with the aim of improving knowledge and procedures.

Information assets

Information, knowledge and intellectual property are the main assets of the modern organisation. The products and processes that are not reliant on the knowledge worker are quickly automated and soon lose their commercial potential. The ability to store information and knowledge, and utilise it, to inform the organisation's decision-making process is an essential requirement of successful organisations. Organisations must seek to recognise knowledge and maintain the systems and human resources that hold the intellectual property, ensuring the information is exchanged and integrated to realise its potential. Construction operates in a knowledge-based economy.

Knowledge storage and transfer

Egbu (2000) identified a number of ways in which knowledge management practices can be transferred and embedded into organisational practice, these include:

- *The creation of knowledge teams*   The staff from all disciplines form teams and work together to develop or improve methods and processes (this also applies to knowledge groups).
- *Shareware*   The provision of information platforms, occasions, events and locations that encourage knowledge exchange.
- *The introduction of knowledge webs*   A formation of a network of experts and communities of practice who collaborate across divisions and strategic business units.
- *The establishment of intellectual capital teams*   The team roles include the identification, storage and auditing of intangible assets such as knowledge.
- *The provision of collaborative technologies*   Use of intranets or groupware allowing rapid information access in remote locations.
- *Establish good practice*   Definition and communication of knowledge-performance behaviours.
- *Embed in organisational policy*   Make knowledge performance and information exchange company policy.
- *Recognise knowledge workers*   Identify key knowledge workers and knowledge-performance positions (and exploit and reward their talents).
- *Reward knowledge-sharing behaviour*   Provide incentives for good knowledge-management practice.
- *Eradicate poor knowledge management*   Take action against those who do not use good knowledge-management practice.

Accuracy

Accuracy of the information communicated to others in the process is an essential requirement. Accuracy and consistency of the words and symbols used is paramount to mutual understanding. The use of common symbols, processes and tests, such as those established by International and British standards, is recommended.

These institutions provide practical guidance on common methods and processes. Education, training and experience will have considerable impact on our ability to communicate information that can be interpreted with high levels of accuracy.

Language

Communication is essentially a social activity, the sharing of information and the sharing of experiences, which is dependent upon the communicators understanding the rules of communication. Speech, writing and drawing are obvious modes of communication, but so too is body language which can convey more subtle, and rarely recorded, understanding. Communication performs a much more complex task than simply transmitting information: it involves language. Language is central to all social activity and involves abstract notions, actions and events removed in time and space, with subtle shades of meaning and logical distinctions that depend on people sharing a complex and symbolic representational system (Potter & Wetherell 1987). Interaction between construction professionals will, to a lesser or greater extent, be dependent on the language and codes used and how they are received and interpreted.

Communication is a vital characteristic of human societies. It is also of vital importance to everyone involved in building (social intercourse). Designers have to be good at drawing, but they must also be able to present their ideas in a manner that those commissioning the design are able to comprehend, i.e. designers must be able to demonstrate the value of their design and the value of the services they offer. Failure to articulate the importance of design and the value of their contribution to the building process in addition to future building users may lead to reduced work and reduced fees. Similar arguments can be put for other consultants and contractors. What we are saying here is that it is of little use being very good at our job if we are unable to articulate and communicate our contribution to those paying for the service. Marketing is a very important part of the equation and must continue throughout the project. Communicating design awareness throughout the different stages of the project is not only an important skill, it is vital to the effective transmission of design intent from design to finished building.

Diverse languages

When architects (and authors of architectural books) talk about communication they are frequently found to be talking about the way in which their work (both their design drawings and the completed building) communicates with the reader. Architects talk of 'reading' buildings and the way buildings 'communicate' with their users and viewers. This is an important and necessary part of understanding architecture although it is unusual to find other members of the construction sector adopting this interpretation of communication. Here, we are concerned with the way in which participants communicate and the various languages employed to achieve mutual understanding.

Construction professionals enter into communication with diverse perceptions, attitudes and values. For example, it has been suggested that architects and construction managers have a limited shared social reality and their ability to communicate is restricted by this situation (Brownell *et al.* 1997). People trained in particular fields have their own vocabulary and language. Words can have very different meanings within fields of specialisation, yet people do not usually define the words that they are using. Stretton (1981) acknowledged that these aspects constitute a major barrier to effective communication. He suggested that before encoding messages the sender should send the message in a language that is acceptable to the receiver, if necessary explaining and developing an understanding of associated issues (which are necessary to create understanding), before the main event is discussed. Such actions are aimed at supporting the receiver of the message, i.e. the communicator is attempting to identify the language and style that are acceptable to the receiver. A problem of identifying the most appropriate form, language and style of communication is based on our perceptions. These are personal (and rarely shared) and so we can only guess at our colleagues' perception. For designers, emphasis is on talking to clients in a language that they understand (one specific to their organisational setting), which will be different from the language used to communicate with other consultants, contractors and building users.

In addition to the complexities of language used between different professional groups we need to mention regional dialect and national languages. Regional dialect can be such that two English-speaking people find it difficult to understand one another. For those working in countries where a language is spoken that varies from their native tongue then communication requires considerably more effort if we are to be understood. With the relatively free movement of labour and the use of foreign labour to reduce costs we should anticipate and hence make allowances for some communication difficulties. More subtle differences can be found in the American, Canadian and Antipodean use of the English language. Here dissimilar meanings attached to words can cause more problems simply because we tend to take for granted the fact that we all communicate in 'English'.

Towards a common language

Arguments for the development of a common language and shared values have surfaced from time to time in construction. With increased emphasis on the improvement of the construction process and the constructed product, the desire for a common language is topical once again. While this may be an altruistic aim, there are a number of closely related factors that deserve some attention first. Namely, the issues of professional roles and status, educational norms and codification and shared values.

Professional roles, status and role expectation

Participants to construction will differ in the type and level of education they have received, their professional values and their knowledge base, which will colour their interaction. Contrary backgrounds, education and training can lead to different perceptions of what is of greatest importance to the project at different times. The cultural differences not only distinguish organisations and individuals, but also affect their primary goals (motivation). Personal divergence between professionals, for example architects and construction managers, can lead to conflict. The historical and professional differences have led to different perceptions of social status and role definition (e.g. Higgin & Jessop 1965, Bowley 1966, Faulkner & Day 1986). In addition to these perceptions the professional institutions have identified procedures and rules of engagement for their members to follow which may influence how different professionals conduct themselves. Gameson (1992) suggests that before researchers can examine interaction between professionals consideration must be given to how each profession has developed, the background to their development, their education, training and social status. There are clusters of perceptions that surround each profession establishing certain expectations of how a professional will behave in certain situations. Handy (1981) suggests that there are traditions that have been established over time that shape the role of the individual in their work situation.

Education

There are considerable variations in the length, structure and content of construction-related higher education courses. Many of the differences in the courses can be attributed to the demands of the profession and professional body that accredits or endorses a particular course. For example, architects still have the longest period of formal education within the industry: seven years to qualify, compared with five years for the majority of the other participants. Another contributing factor is the way in which professionals are taught, in particular the continued reluctance of architects to be educated alongside their future work colleagues, namely the architectural technologists, engineers, surveyors, project and construction managers: this is essentially the separation of design from production in education. One argument associated with a need for greater shared values and hence a more appropriate and common language is that for a common education in building for all disciplines before going on to specialise. It is a powerful argument, but one that has been resisted to date. The problem here is associated with professional institutions' concerns about loss of identity and loss of power, and with it the loss of specialised knowledge and the loss of diversity.

Codification and shared values

Professionals have their own 'special' mystical language, for architects the language is expressed in graphical format, images readily accessible to others who share the common language (other architects), but often quite inaccessible to those not trained in architecture. In architectural education, students are taught how to design and draw, and although a lot of time is dedicated to the pursuit of design excellence, by comparison, little time is spent considering the receiver's perception of the codified message. For example, architectural details contain highly codified information, and are quite unintelligible to the uninitiated. A similar observation could be made of other professionals. The degree of codification is obviously linked to training and professional values. Shared values may well exist at certain times in a project's life, but it would be unrealistic to hope that all contributors to a project would want to have, or were capable of developing, shared values. Projects tend to be more successful when harnessing and exploiting the different values that individuals and different organisations bring with them, i.e. there is considerable strength in diversity. If values are divergent and many would argue must be distinctive, how can a common language be developed? Before answering the question we first have to address the cultural context of the project team and how this colours interaction.

The cultural context of the project team

The problem with many communication studies is that communication is studied with little attention to culture. Culture influences communication (intrapersonal, interpersonal, intergroup, organisational and political). The problem with studying communication in construction projects is that different individuals are drawn from a variety of educational and cultural backgrounds, thus barriers to effective communication are sure to exist and cannot be ignored. Furthermore, the culture of the social system(s) and networks that form the temporary project organisation will influence how individuals within this system communicate. Design is a participatory process. Each member of the team will have their own agenda, goals, individual values and experiences that may differ from the next individual in the project information chain; this will influence the interaction and participation of individuals, in particular it will influence the efficiency of communication between them. The building industry is notorious for its adversarial behaviour and distrust between different professional groups. At certain times these individuals will meet and interact.

Anyone familiar with the field of geography and plate tectonics will know that when the plates under the earth's crust either collide or separate, friction results, and it is these boundary conditions that problems, such as earthquakes, occur. It is useful to draw a comparison between the science of plate tectonics and the building process because there are a number of distinct boundary conditions where friction and thus ineffective communication are most likely to occur, and they need careful consideration if communication is to be improved (see Chapters 7 and 9). First is the boundary between the sponsor of the building project, usually referred to as the client, and the designer, usually an architect. In modern parlance this is known as the briefing stage (pre-contract) during which client and designer communicate with one another until a design brief has been produced. Second is the boundary between designer and contractor. Once the designer has manipulated design knowledge into a meaningful design, it has to be communicated to the contractor, usually through drawings, models, specifications and schedules. Third is the boundary between contracts manager and tradespeople who actually assemble components, systems and products on site to form the finished building. These are the major boundaries where faults can and do occur. But there are many smaller boundaries, for example within the design office where it is common for a senior member of the firm to take the client's brief and communicate it to less senior members of the office who will work on the project. These, too, can lead to barriers in the flow of information. In many respects the issue of communication breakdown, gatekeeping behaviour and communication network has been a matter of conjecture, with little research in the field addressing such behaviour, which admittedly, is difficult to observe. Such studies have taken place with the relatively broad field of communication and psychology and this book draws on some of the more relevant work in these fields, together with the authors' own work in this area.

The ability to share information is critical to expert knowledge systems and information management systems. While free access to information is possible within an organisation (although the organisation's culture may inhibit this) access becomes problematical when looked at in terms of the temporary project environment. Some participants may keep information back as a means of gaining some form of advantage (i.e. acting as a gatekeeper) and partly because many of the project team members may well be competing for the same market segments, thus security and policing of the system become overriding, and restricting, factors. In practice the extent of 'managed information' may be (very) limited. In short, the ability to share information is a complex technical and social problem that the construction sector is still struggling to come to terms with.

An appropriate language

Returning to the question we posed earlier – Do we need a common language? – the answer will, of course depend on the reader's own values and beliefs; however, we would argue that it is more important to understand communication as it relates to a particular set of circumstances. Improving communication across intercultural, organisational and project boundaries must be central to improving the quality and enjoyment of the design and construction process and hence to the quality of the constructed works. Selecting an appropriate language for a particular situation is key to this strategy. Spending too much effort on creating a common language may do little other than to distract from the richness and potential embodied in the cultural diversity of the construction project.

A time and a place

Communication across cultural and procedural boundaries takes a degree of skill and effort, as do the use of an appropriate language and the choice of appropriate communication media and communication channels. Recognising this, however, is not enough. We must also consider the interrelated factors of time and place because both factors will colour the communication process.

Time

It takes time to develop designs and time for the receivers of the information to understand what is communicated to them. One of the problems with the speed of digital transmission is for people to think that just because the information has been sent within less than a second from one office to another, that the information is instantly read and understood. This is not the case. Time is required to understand the various aspects of a particular project (an observation that also holds true for projects with a high proportion of prefabrication) before acting on that information. We all complain of trying to do too much in too little time and it is critical that the production and subsequent use of information are programmed to allow the sender time to consider the needs of the receiver. Programmes of work should also allow the receiver adequate time to read and understand the information, and of course provide the opportunity to ask questions to aid understanding. This takes on even greater significance when information packages are phased to suit fast-track construction programmes. This works both ways. Conditions in construction contracts stipulate a certain period of time for the designer to respond to requests for further information and/or clarification of the information provided from the construction manager. It is to our collective benefit that we all try to abide by such conditions.

Place

Communication does not take place in a vacuum. The environment in which communication occurs may either help or hinder the process. By communication environment we mean both the physical surroundings, be it a warm office or a windswept site, or the media environment in which the information is contained, and the perceived environment, be it supportive or defensive.

Physical environment

How we feel about our physical surroundings will influence how we communicate. It will come as no surprise to find that familiar and comfortable environments encourage openness while unfamiliar and uncomfortable environments tend to make us defensive. Construction sites can be rather daunting environments for the uninitiated, never the same on consecutive days and rarely comfortable. How, for example, are we to conduct a serious conversation on a site with noisy and dangerous activities taking place around us. Add in some wind and rain and the tendency is to make snap decisions rather than discuss issues in adequate depth. The level of familiarity and the comfort afforded by it will affect the time we allow for communication and hence will influence our decision-making behaviour. The environment in which information is communicated (media) should also be considered. It is clear that some people are better able to use paper-based systems than electronic systems, and vice versa.

Perceived environment

When we enter a social environment we will make certain assumptions about what we are likely to experience. Assumptions will be made about the types of people we will engage with and the way they act in a certain environment, which is based on our previous experience of similar events. Our perceptions will result in emotions and feelings that are used to prepare the way we engage with and respond to others, i.e. they affect the nature of interaction. For example, if a contractor holds a strong perception of how architects behave, then the contractor may use this information to develop a communication strategy for dealing with the stereotypical architect. Regardless of the architect's behaviour, the contractor will make assumptions that will result in the adoption of communication behaviour that will affect the relationship with the architect. Apart from reinforcing stereotypical images of other professionals, the danger is that we see and hear what we want to hear, not what is being done and said. We must be aware that we behave in this manner and then make an effort to be a little more responsive to the actions of others.

4 The dynamics of communication

The application of communication models to construction is possible and has been attempted on a few occasions. In all cases such application has been accompanied by a series of caveats that attempt to deal with the peculiarities of construction and hence a number of generalisations and assumptions are made to enable comparisons to be drawn. Before we can apply interpersonal communication models we need a full understanding of the fundamentals of human communication. This chapter provides an overview of the main communication models that may be applied to construction and summarises the different levels of communication.

The development of communication research

The study of communication has been documented for over 2500 years with scholars such as Plato and Aristotle conducting research into verbal messages and their civic affairs (Philipsen & Albrecht 1997). However, the body of sustained empirical research into communication has developed during the later half of the twentieth century, a period that has been described as the 'age of communication' Rogers 1986: a period in which communication science has flourished, fuelled in part by massive advances in information technology. The history of communication science is recorded elsewhere (e.g. Rogers 1986) and does not need repeating in any great detail here; however, it is important to recognise that, like construction, communication science is not a coherent field, it has differing roots and sub-fields, hence terms vary in their use and meaning. For example, diffusion of innovations work has developed largely in isolation from work into gatekeeping behaviour, despite the clear synergy between the two. There are a number of well-known pioneers of communication research who set in train the momentum for later empirical work. Gabriel Tarde's (1903) early work can still be traced in the diffusion of innovations literature, and Georg Simmel introduced the theory of communication networks with his book *The Web of Group-Affiliations* (1922, translated into English in 1946). However, it is the work of Shannon and Weaver that is most widely known.

Early communication models

Shannon and Weaver's *The Mathematical Theory of Communication* (1949) provided a model of communication supported by mathematical theory. Their model was represented by a very simple diagram (Figure 4.1) which resulted in its universal adoption by communication scholars and led to a linear approach to human communication studies. Although the model has been criticised for its simplicity: it was not as linear as their (misleading) diagrammatic representation suggests because they recognised that the encoding of messages into signals and their later decoding was a subjective process (Rogers 1986) and they borrowed the word 'entropy' from physics to help explain the degree of uncertainty in the system being studied. The

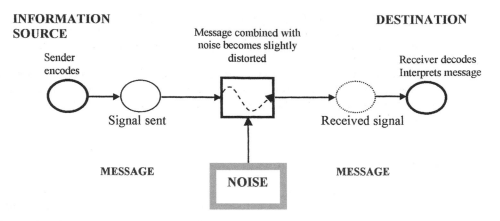

Figure 4.1 A 'linear' model of the communication process. Source: Adapted from Shannon & Weaver 1949.

concept of feedback was not explicit in their work, furthermore the model was difficult to test because it had too many component parts. This led to a simplification of the model into a Sender–Message–Channel–Receiver model that was more applicable to studies of mass communication. The simplified interpretation of the model is that for information to be communicated it must be:

- encoded by the sender;
- transmitted;
- resistant to the effects of distortion due to noise. During transfer information encounters noise. Noise is any signal or other occurrence that can distort or interfere with the transfer and interpretation of data;
- decoded by the receiver. When the signal reaches its destination, the receiver attempts to interpret and understand the message, i.e. they attempt to decode it.

Definitions

Defining what we mean by communication can prove to be difficult. A survey undertaken by Dance and Larson in 1972 found 126 definitions of the word, since which time the number of definitions has increased (Trenholm & Jenson 1995). Writing in 1990 Fiske noted that communication is one of those human activities that everyone recognises, but few can define satisfactorily. Although the roots of communication theory go back to the mechanistic Sender–Message–Channel–Receiver model, in which information is transmitted from sender to receiver (implying control over the process in which the power rests with the sender of the message), the model has been adapted to recognise that communication is a two-way process. More recent work has moved to a shared perception model in which each person is a 'participant', rather than a 'sender' or 'receiver' (Rogers 1986). One of the more robust definitions is provided by Rogers and Kincaid (1981: 63) who define communication as 'a process in which the participants create and share information with one another in order to reach mutual understanding'.

From this definition it follows that 'information' is exchanged in the communication process as participants create meanings. A more extensive and pertinent explanation is provided by Tubbs and Moss (1981), the main components being:

- The creation of meaning between two or more people;
- The essence of communication being to send, place, exhibit or manifest a message, signal, code, movement or other stimulus which means something to the receiver;
- Information communicated will not mean the same to the sender, but invoke a reaction, manifest a thought that has relevance to both receiver and sender;
- The relevance of the communication need not be the same to the sender and the receiver.

The process of communication

Communication is used by people to gain control over their social and physical environment and the importance of communication in a business setting cannot be understated. Exerting a positive influence on our business environment can be achieved through effective communication, but to do so requires a thorough understanding of communication. The two main approaches to the subject are the 'process' method and the 'semiotic' method (Fiske 1990), summarised below.

Process

The process method sees communication as the transmission of messages, through which one person (or organisation) seeks to influence the behaviour or state of mind of the other. When the outcome of the communication process is less than expected it is viewed as a failure. This approach to communication is drawn primarily from the fields of psychology and sociology and is concerned with how:

- Senders and receivers encode and decode messages
- Channels and media are used to transmit messages
- Efficient and accurate the communication act was.

Semiotic

The semiotic method sees communication as the production and exchange of meanings, primarily concerned with how messages are used to manifest meaning. The difference between this school of thought and the process school is that mis-understandings are not necessarily considered to be evidence of communication failure, rather evidence of cultural differences. Semiotics is the science of signs and meanings and draws upon linguistics and the arts. The importance of signs and symbols is highlighted in this work, and overt behaviour may be less important than hidden messages. It is concerned with how:

- information, meanings and feelings are shared by people;
- verbal and non-verbal messages are produced, processed and delivered (exchanged);
- messages affect those who receive them.

The different views are articulated in Fiske's work, which suggests that both deserve consideration. Given the issues raised in earlier chapters, such advice appears to be timely. So, for the purposes of this book, we will view communication as a process in which messages are used to manifest meanings, thus combining both schools of thought. It follows that if we view communication as a process, the natural thing to do is to apply a suitable model to allow us to study it. However, we have made the point above that there are a number of models and their suitability depends on the subject of the study. Furthermore, diagrammatic representation can, as we have seen, be misleading. So rather than use a model, we have outlined the main elements and discuss them below in relation to construction. The communication process starts with the sender's need to transmit a message.

The sender

The sender, sometimes referred to as the information source, is a person who transmits an initial signal (message). This message is encoded into a suitable communication medium – words, drawings, gestures, etc. – before it is sent. A good example would be the communication of information from the architect's office to the construction site. As originator and sender, the designer would choose the media that he or she feels are most suited for conveying the intended message; this may be a single drawing or a series of drawings supported by notes, schedules and written specifications. This is then transmitted to the contractor, who, we hope, is able to decode the meaning conveyed in the message. We tend to assume that just because we understand drawings and schedules that the receiver will have the same understanding. This is misleading because the information selected for transmission and the media chosen by the sender may not necessarily be in a form that the receiver can easily understand, so the potential for misunderstanding is always present. The sender must make an effort to anticipate the receiver's needs, and this is difficult if the sender is preparing contract documentation with no idea as to who is going to use it.

The message

The message is an encoded idea that is transmitted in a suitable communication medium through a suitable communication channel. The message may vary in complexity, ranging from a simple drawing to confirm dimensions through to complex construction detail that requires drawings and notes to explain it. We may encode our thoughts into speech, verbal media, or through drawings and letters and non-verbal media. The skill is to encode the message in a way that will ensure full understanding by the receiver. A remarkably simple statement to make but one difficult (some may argue impossible) to achieve in practice because the message will be decoded by someone with different experiences, attitudes and motives to the sender, thus understanding is likely to differ between sender and receiver. The situation is complicated further in construction because it is likely that the receiver will come from a different educational and professional background to the sender, and use a different language of codification to that of the sender. In selecting the appropriate message it will be necessary for the sender to pass the idea through his or her own 'filters' to ensure it has meaning to the receiver before it is codified. This process will 'colour' the message before it is sent.

Communication channels

We tend to communicate in one of two ways, either formally or informally. Formal communication channels are associated with the project's contractual requirements, whereas informal communication channels are seen to lie outside.

Formal communication channels

Communication events that are formalised are in some way structured, e.g. prearranged meetings and management systems. Formal communications are the accepted system of communication within the organisation; they are the official sources of information using prescribed channels. Systems, channels and events are organised and structured by managers in an attempt to ensure that essential information is processed. These systems are usually structured so that information exchange is recorded for future reference, a process facilitated by information technologies. Organisations will adopt a formal structure for communication so that all members, regardless of position, know whom to ask and whom to inform. Attempts will be made to control information so that individuals do not experience information overload and more importantly that they receive information that they need. Construction projects will adopt a formal communication structure as set out in a particular form of contract.

Formal communication is sometimes classified by the direction of movement in relation to an organisation's hierarchy system. Movement can be described as vertical (downward, upward) or horizontal (sideways). Downward communication (top down) is the information that is distributed from management, with higher authority, to the workforce or managers of a lower authority. It generally involves the giving of instructions, dissemination of company documents and safety information, etc. Upward communication (bottom up) involves communication from employees to their managers. Examples being requests for information, the provision of progress reports and feedback on aspects of progress.

Informal communication channels

Informal communication channels are routes of communication other than those identified by the organisation. Differences between informal and formal communication are normally associated with the degree of control. Formal systems of communication are in some way controlled, or organised, whereas informal communication systems are largely unstructured. Informal communication channels emerge through friendships or contacts between individuals who are willing to co-operate. They may be seen as communication shortcuts, unofficial ways of receiving required information, thus avoiding overly bureaucratic channels and/or organisational gatekeepers. Gaining information through informal communication is connected to help-seeking behaviour, which may be used to encourage supportive communication and break down defensive communication, helping to strengthen informal relationships.

Middleton (1996) found informal conversation constituted a key element in multi-disciplinary professional teams. Arguments and discussions enabled greater understanding of events, improving the co-ordination of activities. Middleton's work examined the informal behaviour of 'corridor talk and chit chat' as a mechanism for improvising interim solutions to unexpected problems. Informal conversation enabled the team to maintain up-to-date knowledge. Discussions involving activities of the team provided a forum for the development of common knowledge or 'working intelligence'. Pietroforte (1997) also found that informal communication was necessary to make construction contracts work. Furthermore, informal conversations may aid problem-solving and decision-making in meetings. Hastings (1998) suggests that it is important to build in 'informal social time' before and during meetings, it is often in these social interactions that bonds are formed and real issues get discussed. However, one must always be aware of business interactions becoming too informal.

Communication media

The sender of a message has a wide choice of media from which to choose. The benefits of one over another will depend upon the message being conveyed and the communication channel through which it will pass. This is explored in detail in Chapter 10.

The receiver

The receiver's understanding of the message conveyed will be based on their perception and understanding of the information at a particular point in time. The encoded message will be interpreted by the receiver's sensory organs and decoded into meaningful ideas. This should allow the receiver to understand and share the sender's intentions, however, it is a complex process and sometimes the sender's message may be distorted when the receiver attempts to make sense of it. In the worst case the idea may never be shared between sender and receiver; at best, the receiver may recognise the need for further information in order to understand the sender's intentions. If communication is one-way there may be little opportunity for the receiver to ask questions to reduce their uncertainty. Fortunately, in the majority of cases the opportunity for two-way communication exists and the participants have the chance to share messages by asking questions (see the section on feedback, below).

It has been argued that we can never achieve common understanding in its truest sense. Differences in experience, education, background and ability to process information combine so that our level of understanding may be similar, but never identical. Figure 4.2 illustrates how information can be shared to bring about the overlap in understanding.

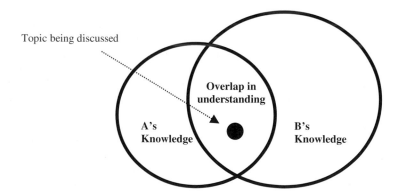

Figure 4.2   Levels of knowledge and experiences of topics being discussed.

Parties A and B have considerable overlap in their understanding of issues related to the topic being discussed; however, party B has far more knowledge and experience specifically related to the topic. Even though B has a greater knowledge base, 'A' has experienced a number of situations that B has not. To increase their understanding and develop an appreciation for each party's knowledge the parties must enter into conversation, discover common ground, then develop and build a greater level of congruent understanding.

Receiver's selective attention

So far we have assumed that the receiver is expecting and will respond to the message, but this is not always the case. Individuals will exert a degree of selective exposure to messages, even formal ones, and so we need to address the issue of selective exposure here. This is a natural and necessary action to limit the negative effects of information overload. Some events are more salient than others, for example if a manager is under pressure to hit a deadline with a specific task, any information that helps him or her to achieve the task will be given greater consideration than data that can be dealt with later. Such selective attention can cause us to miss important information if it is not presented in a way that makes us focus on the message.

Another question that the receiver will be asking is whether the message is relevant to their particular needs at a particular point in time, i.e. is the message timely? The answer will be made at both a subconscious and conscious level. Factors that will affect a receivers' effort to understand and process the information will depend on whether the receiver:

- Was expecting the information
- Is likely to be affected by the information
- Has prior knowledge of circumstances surrounding the information
- Has background understanding, and is aware of links with other information
- Is able to recognise when information is missing
- Is able to request further information and can recognise when information is still incomplete
- Is able to identify the importance of the message. People often use a range of emotions, language, signals and movements to identify and separate important and unimportant information.

Feedback

In the majority of cases the recipient of the message has an opportunity to ask the sender for additional information or for clarification. This is usually referred to as feedback in communication models. The important point to make here is that if the receiver does not fully understand the message he or she may well use a different communication medium in an attempt to reduce uncertainty. For example, if a vital dimension is missing from a drawing the site manager is likely to telephone the designer's office for the information (usually because it is urgently required). The reply may well be verbal, followed by confirmation in writing. During face-to-face interaction, feedback and communication signals are exchanged simultaneously. As a person sends a signal, the receiver instantly responds with subconscious, and sometimes conscious, non-verbal signals. Such exchanges provide clues regarding the effectiveness of the initial message – whether the message is being received, understood, agreed or disagreed with, and whether further explanation is necessary.

Time

Communication is a process and therefore we cannot ignore the influence of time. Throughout the design and construction phases of a construction project there are time pressures. These pressures are contained in programme deadlines, e.g. deadlines for sending information to other consultants, deadlines for sending the contract documentation to the main contractor, deadlines for practical completion and hand-over of the building to the client. Therefore, at certain times there will be pressure on individuals to produce information quickly and the possibility of sending incomplete messages (which may be incomprehensible to the receiver) is a real threat, thus triggering a request for further information. Research has shown that imposed time constraints generally result in people opting for simpler decision-making strategies (Edland & Svenson 1993, Ordonez & Benson 1997). It is not unreasonable, therefore, to expect individuals to avoid seeking further information when faced with a deadline, which may complicate decision-making.

Communication models

Disregarding problems of accessing sensitive business environments there are inherent problems of modelling communication. Although early models of communication were simplistic and linear, more recent theories of communication and cognition, such as Sperber and Wilson's relevance theory (1986) are quite complex and dynamic. Indeed some theories, by their author's own admission, are identified as difficult to understand unless the reader already has knowledge of a specialist topic (Masnikosa 1999). Even if it is not necessary to have an understanding of a specialist area prior to reading the theories, the study of human communication is a complex phenomenon. The interaction of multiple parties subject to the psychological, social and contextual influences associated with group communication makes it one of the most difficult objects of study in the human sciences (Hirokawa & Poole 1996). Researchers of communication in construction are faced with the study of a developing and changing environment, comprising various professionals embraced in dynamic communication responding to the project needs. Figure 4.3 identifies a number of factors which should be considered when investigating organisational communication.

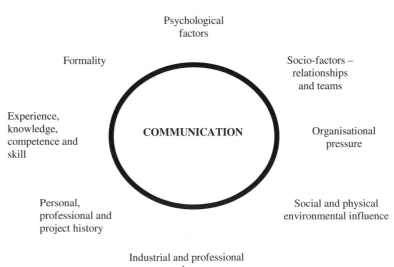

Figure 4.3   Some factors that affect organisational communication.

Models

Many undergraduate textbooks opt for the simplicity of the Shannon and Weaver (1949) model, which characterises communication as relationships between input and output. Their model was developed to determine the maximum amount of information that could be conveyed along a single cable in their work on telephone exchange systems. Although they claim that the model is widely applicable to the whole question of human communication, not all agree. Coded models may be too simplistic, they fail to explain what communication achieves and ignore feedback. However, the coded models still provide a model of communication that is popular in construction publications (e.g. Calvert *et al.* 1995) and has been used to ground research observations of aspects of the construction process (e.g. Bowen 1993, 1995).

Two models of communication that emerge from the many available, approach the issues of communication in a more comprehensive manner. They are Feldberg's (1975) model of human communication and relevance theory (Sperber & Wilson 1986).

Feldberg's model

Feldberg's (1975) model identifies issues that affect the people involved in communication. This model is a significant development on Shannon and Weaver's encoding/decoding model, producing a much more comprehensive theory. The model is linear and mechanistic, being presented in four stages (Figure 4.4).

The first stage assumes a two-person communication process, with parties performing the role of either the sender of signals and messages, or the receiver of signals and messages. The identified components at this stage include sender, receiver, the message communicated, the medium used, the individual's expectations, their reaction to the message or signal, and the result, direction of the message and the content. The sender's expectation, which can be seen as the anticipated result, and the receiver's reaction, the actual result, is viewed as a combined result. The degrees to which the expectations of the sender conform to the reaction are said

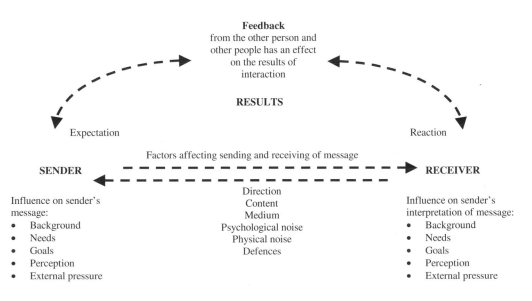

Figure 4.4 Model of human communication. Source: Adapted from Feldberg 1975.

to depend on a number of factors. These factors include the direction of communication, the medium used and the content of the message. The sender should select the most effective medium and content in relation to the receiver and the distance the receiver is from the sender.

The second stage evaluates why the sender's expectations and receiver's reactions are incongruent, the principal factors being external pressures, personal factors, physical noise affecting the signal, defence mechanisms and psychological noise. It is claimed that these factors cause different perceptions of reality, which may cause incongruent understanding, expectations and reaction.

The third stage identifies mechanisms for evaluating the relative success of communication. Evaluative mechanisms are based on the feedback from the receiver or others. The sender evaluates the feedback according to the source and the expectations of the sender.

The final stage is the reaction communicated by the receiver. The process continues and is reversed until the desired result from the communication process is achieved. The process will continue until the communication is terminated. The transfer of any message encounters noise (psychological noise and physical noise) and defences, and is affected by the direction, content, medium and feedback. To summarise Feldberg's model (1975), the main issues surrounding the sender and receiver include:

- Needs
- Perceptions
- Goals
- Background
- External pressures
- Expectations and reactions
- Feedback (from more than one source and possibly through different media at different timescales).

Although this model helps to illustrate what is being processed, it does not explain how it is processed or on what basis the information is assembled to achieve understanding. Nor does it explain cognition. The analysis of language and communication should go beyond the basic building blocks of words and sentences. An understanding of cognition is essential to understand how communication is taking place, for which a background of shared social reality must exist. It is this common ground and its subsequent development that makes communication possible. For the architect and construction manager to communicate effectively they must have an inclination of what the other person might understand: assumptions about their knowledge and experience needs to be made. Once the speaker is aware that the addressee has an understanding of a situation they may be able to communicate on that subject. Where there is a lack of congruent understanding the speaker has to provide an infrastructure on to which the new information can fit and be understood. During interaction, communicators develop a theoretical framework of the other's knowledge. At the point when a person recognises that specific information is understood, assumptions are then made about related information. Needless to say, a decrease in an individual's ability to use the other's knowledge limits communication considerably (Brownell *et al.* 1997).

Relevance theory

Relevance theory is important in the context of this book because it addresses communication from a human cognition viewpoint. The theory is that the human brain is concerned with achieving the greatest possible cognitive effect with the smallest processing effort, thus an individual focuses his or her efforts on sending and receiving relevant information (Sperber & Wilson 1986). The focus is on human cognition, the understanding and processing of one's physical environment, addressing the association of any relevant meaning of any message or signal in any form. The theory of relevance discounts code models on the basis that the complex nature of communication causes difficulties in arguing the existence of a comprehensive code which covers all aspects of communication and can place those in the mind of the sender and receiver.

Communication is dependent on cognitive abilities. Cognitive ability includes the processing of sensory information into experiences and probable experiences. Facts are acquired, constructing knowledge of facts, reinforced assumptions, assumptions with other relevant assumptions, enabling the ability to become aware of further facts. Relevance also involves what is known as selective attention; we filter out a lot of sensory information from our environment, mainly to avoid overloading our cognitive resources. Many other studies also agree with the notion of selective information processing (Huseman 1977, Glass & Holyoak 1986, Burgoon *et al*. 1994, Price 1996). What we take in will be relevant to our current interests and will help to focus our attention. This focus may change as something more relevant takes precedence.

Knowledge of past events and experiences enables or prevents us from processing more information than necessary. At any one time a human is exposed to many different sensory experiences, processed into environmental information, some of which are much more salient than others. The information that is more salient holds a higher priority, and is more relevant. It follows that incoming information is built on existing information and experiences of a related nature. Thus, following an experience, related information is accessed. Information comes in many forms, it is not necessary to have a complete mental representation of something to know facts about it (Sperber & Wilson 1986).

Cognitive environments are constructed of facts perceived, or inferred, and made into assumptions. An individual's total cognitive environment comprises all the facts that are manifest to him/her. Any observation or sensory experience can be manifest but is not necessarily known or assumed. The same facts and assumptions may manifest in two different people; however, this does not mean that they make the same assumptions, although they may be capable of doing so. If the cognitive environment is a set of assumptions, then communication is aimed at stimulating assumptions in the mind of the receiver. Relevance theory is based on the interconnection of old and new information. If we are unable to draw a significant link between what we already know and other information, it will not be understood.

Issues for consideration

There are a number of aspects that we feel should be considered when using and/or developing models of communication. They are:

- Relevance
- Relationship and situation
- Selective attention
- Psychological and physical noise
- Nature of communication – task-based or social
- Media used
- Information load, emotional tone

- Ability of participants to process information, skills, knowledge, attitudes, culture
- Motives, thoughts, beliefs, goals
- Message, signal, feedback (non-verbal and verbal)
- Environmental stimuli
- Level of communication.

Levels of communication

Communication (and the absence of communication) is used by individuals and organisations to achieve a number of objectives. First, communication channels are used to inform, transmitting information from sender to receiver. Second, communication channels are used to both establish and maintain relationships. Third, communication may be used as a tool to influence individuals' behaviour. Thus communication forms the link between human behaviour and management: 'management through communication' to use Roodman and Roodman's (1973) words. Communication holds a central position within organisations. The manner in which an organisation is structured and operates, just like its effectiveness, will be determined by the communication techniques employed (Barnard 1938). A similar statement can be made for the success of the building project.

Human communication may be divided into four levels (Kreps 1989):

- Intrapersonal communication enables an individual to process information.
- Interpersonal communication enables individuals to establish and maintain relationships.
- Small-group communication enables members of work groups to co-ordinate activities.
- Multi-group communication enables different work groups to co-ordinate their efforts.

Intrapersonal communication deals with the cognitive process of an individual, investigating how they process and build their thoughts, assumptions, knowledge and beliefs. At the other end of the scale is mass communication. This focuses on communication media that have the ability to communicate to many individuals at the same time, e.g. television, worldwide web, newspapers and other publications. Interpersonal communication is important because it is at this level in the communication hierarchy that relationships are established and through which individuals co-orient their behaviours towards common goals. Hence interpersonal communication is crucial to co-orientation and the ability to organise (Kreps 1989) and is fundamental to effective communication within small groups (e.g. within the office) and multi-groups (e.g. within the temporary project team).

Groups will develop a structure over time based on (1) power, status and authority, (2) individual roles within the group and (3) the degree to which individuals like or dislike each other. Structure and communication are irretrievably interlinked (Hartley 1997). Communication studies have identified the importance of network structures (discussed later) and the role of individuals within networks who may act as gatekeepers to the flow of information during interpersonal communication (see Chapter 5).

Table 4.1 provides a brief explanation of the different levels of communication. This provides a simple insight to the relationship between interpersonal and group communication and the other levels of communication.

Table 4.1   Levels of communication

Process	Number of people involved
Intrapersonal communication Internal communication process (cognition) includes the manifestation of information in the brain, which is understandable to us. Intrapersonal communication would also include the knowledge that another person is able to process information (relevance), which provides the initiator of communication with the knowledge that s/he can communicate with a person.	Only one person involved. It is the thought process of one person either when they are alone or communicating with others. Intracommunication may be used when one person makes a decision. As there is only one person involved some scholars do not view intrapersonal communication as a communication process.
Interpersonal communication Communication directly between two people, enables individuals to establish and maintain relationships. It involves the transfer of signals and messages that manifest themselves in both parties to communication. Intrapersonal and interpersonal communication functions enable information to be processed and joint decisions to be made.	Generally two people (dyad) – more than two people may be considered to be a group – the significant difference being that in interpersonal communication, between two people, the message is intended only for one receiver. Some scholars do not differentiate between interpersonal communication on a one-to-one basis and that of a small group, yet there can be many differences in the nature of interaction.
Group communication Messages are sent to the group, they may be presented in a way that addresses the whole group or individuals within the group. Message may be interpreted, processed, by individuals within the group in different ways. Terminology and language may be specific to the group, each group will have its own culture and norms.	More than two people but limited to a single group of people. Communicators may address the whole group or individuals within the group. Even when individuals communicate within a group this action will be communicated to the rest of the group.
Multi-group communication A person or group communicates a message to a number of different groups or sub-groups, the response to the message may be different depending on the group's motivations and norms.	Although communication of this nature targets a number of groups or sub-groups there is an element that the messages are largely contained within the specific groups, e.g. departments within an organisation.
Mass communication Messages are sent through media – radio, television, and newspapers – or to large audiences. Individuals and groups of people receiving the message may attach different meanings to it depending on their culture and norms.	Little control of who and how many receive the message, groups can be targeted, e.g. television viewing at a particular time. Professional journals are used to send information to their profession.

Further reading

Goleman, D. (1996) *Emotional Intelligence*, Bloomsbury, London.

Hargie, O.D.W., Dickson, D. & Tourish, D. (1999) *Communication in Management*, Gower, Hampshire.

LeDoux, J. (1998) *The Emotional Brain*, Phoenix, New York.

5 Interpersonal communication

An essential skill of all professionals is the ability to express themselves clearly and concisely. Interpersonal, or face-to-face, communication is one of the most common forms of communication, be it between designers in the office or between operatives on site. It is used to gain more information, to question areas of uncertainty and to communicate decisions to others. Here we provide an overview of interpersonal communication and look at the associated issues of influence and control, persuasion and defensive communication. Making informed decisions and the issue of conflict are also addressed.

Professional interaction

In the previous chapter we identified the different levels of communication involved in professional interaction. These are now considered in more detail, starting with intrapersonal communication.

Intracommunication

Intrapersonal communication (intracommunication) is a term used to describe the thinking process that occurs within and to the self. These 'conversations' are the thought processes and reflective thinking that occurs within our minds and is seen as the root of other classifications of communication. As such it is an important element in our decision-making process. Intracommunication is the most basic level of communication and is essential for all levels of interaction. It is used for processing data, for encoding prior to sending a message and for the decoding and processing that occur when receiving a message (Kreps 1989). The categorisation does not sit too easily in the definition of communication as a process involving two or more people. For example, Burgoon *et al.* (1994) state that it is obvious that people think, reflect, and have internal dialogues with themselves, and argue that intrapersonal communication is too broad a definition to consider it to be a separate act of communication. Although some scholars want to separate studies of intrapersonal behaviour from the communication process, practical benefits have been gained from investigations in this area. Developments in information systems and computers have benefited from the studies of intrapersonal communication. Intrapersonal communication is also important for investigating how we process information that comes from visual and audio stimuli other than those created by humans. Unfortunately, this 'black box' thinking and dialogue with oneself is hidden from the observer.

Interpersonal communication

Interpersonal communication is the direct interaction between two people, and in contrast to intrapersonal communication this transaction can be observed.

Communication between two people is sometimes referred to as a 'dyad', which is the smallest unit of human interaction, being a microcosm of group dynamics. Research in this area is usually concerned with face-to-face communication, the verbal and visual interaction between two people, although it has been extended to include verbal exchanges by telephone and the use of non-verbal media such as letters, faxes, email and drawings. The term interpersonal communication has also been extended to cover the exchange between more than two people.

Interpersonal relationships are developed in response to the interpersonal behaviour of each individual, their responses being a reaction to the actions of others (Kreps 1989). This behavioural rule is termed the 'norm of reciprocity', where an individual formulates their actions in a particular way depending on how others behave, and through which relationships develop.

Metacommunication

Metacommunication describes the signals that are exchanged during communication, which tell us whether what we are saying and doing is considered to be correct, which is only possible through feedback. Feedback signals are directly related to the relationship, group or context in which communication is taking place and provide information on norms, social rules and the politics of interaction. The role of metacommunication is to inform others of the correct rules of communication and behaviour, with messages contained in body language, facial expressions, remarks, emotion, verbal pitch and pauses in speech, etc. It follows that metacommunication is an important element in the development of relationships. Indeed, it is this greater understanding and recognition of mutual expectations that help us to accomplish tasks and realise our joint goals. All of this information can be communicated without explicitly discussing it. To be an effective communicator people must be able to recognise metacommunication and be able to learn the appropriate ways of communicating in different relationships and environments, i.e. they need to learn the language appropriate to a particular situation at a particular time. As relationships develop so do implicit contracts enabling the building of strong relationships.

Face-to-face interpersonal communication

Some researchers limit the study of interpersonal communication to that of direct face-to-face communication, excluding any communication achieved via the use of other means. Others see intercommunication in a wider context, focusing on communication between individuals and examining different settings, media and influences. It is, however, the face-to-face exchanges that are the most intimate and potentially the best means of achieving effective communication. Face-to-face interpersonal communication tends to be spontaneous and with maximum feedback (Trenholm & Jensen 1995), with around 93 per cent of the message sent non-verbally (Richard & Kroeger 1989). Two people in close proximity sending, receiving and processing both verbal and non-verbal stimuli will result in an almost spontaneous interaction, which is usually less guarded and structured than other forms of exchange. Messages given out and received include facial expression, eye movements, dress, body language, physical movement, posture, proximity, smell, verbal information and speed of reaction, etc. It is the close proximity of the exchange between two people that tends to be lost, or certainly reduced, in small groups of three or more people. The fact that there are only two people involved in the exchange means that any statement will be polarised, and while there is the potential to discuss matters in some detail it is this polarisation that can sometimes lead to communication breakdown.

With each interpersonal situation the inherent rules of engagement will differ. These rules are largely unconscious, and draw on metacommunication. Interpersonal communication can be expressed as three different types of interaction, namely, linear, interactional and transactional:

(1) *Linear* A one-way view of communication. The message is sent and received, the only focus is sending the message and inducing a reaction. No thought is given to any feedback that might be sent. It is, however, difficult to accept the linear view because when people do not seem to react to messages, the absence of a response has significance. When one party chooses not to reply or engage in conversation, this may indicate that a person is thinking, considering a proposition or is refusing to continue discussions.

(2) *Interactional* Interactional communication introduces the concept of feedback. Each message sent induces a reaction, i.e. communication is a two-way process. Based on feedback, new messages are sent, thus messages are sent back and forth between the communicators. A person may set out to say one thing, but change the nature of the message in response to the receiver's reaction to the first few words of the sentence.

(3) *Transactional* The third view takes on board the participants who are involved in communication and any stimuli that could cause a reaction during interpersonal communication. Events that happen simultaneously during interpersonal communication will be processed by the individual. While a person is speaking the content of the message may change as the person receiving the message expresses their feelings and understanding by way of facial expression and body language. Information is also processed from any event in the environment that manifests itself in either of those communicating. The transactional theory offers the only comprehensive attempt to build a complete picture of the communication process.

Public and mass communication

Public communication takes place when one person or a group of people addresses a large audience. Lectures, speeches and presentations are forms of public communication. In public communication speakers often assume that they are the only person sending messages; however, feedback from the audience will be a part of the communication process. The audience's non-verbal (and possibly verbal) reactions to the presentation will send messages back to the speaker, who will respond accordingly (for example, changing the pace of delivery, or adjusting the message conveyed).

Mass communication occurs when information is sent to a large, possibly anonymous, audience. Information is usually distributed through specialised communication media, for example the World Wide Web, television, radio, magazines, journals, newspapers, etc., and feedback is indirect. Interpersonal communication plays an important role in how the information communicated via mass media is accepted. When information is conveyed over the radio or television we will react differently depending on whom we are with and our social interaction prior to receiving the message. Equally, the information that we receive via mass communication may change our perception of events and the manner in which we discuss such issues with others. There is a link between the influence of mass communication on interpersonal communication and vice versa.

Communication, power, influence and control

We use communication as a tool to achieve social influence. We use it to develop friendships, find out information, change others' ideas, influence perspectives, win arguments, instruct people to do things and for socialising with others. Whether it is talking to a stranger at a train station or attempting to motivate workers, communication affects and influences the way that we behave. There is always a reason for communication, even if it is subconscious. We are affected by information and affect others by our non-verbal and verbal actions, we constantly process environmental information, picking up clues to enable us to integrate and react to others.

When someone talks to us, we cannot help but be affected by the information received. As soon as we hear someone speak, see their body language and their facial expression, signals manifest themselves in our brain, whether we want them to or not. As sensory information is processed we react, either subconsciously or consciously. We may subconsciously make decisions, consciously attempt not to change our outward expression, consciously consider our options or react without control, for example faint, freeze, shake, sweat, etc. However, the important aspect of communication is that it is affective and powerful. Many of the responses to signals are made subconsciously. Indeed, even speech is a product of the subconscious, when engaging in face-to-face conversation there is insufficient time to consciously process the grammatical structure of sentences. Through previous experiences we develop a repertoire of phrases, sentences, words, reactions that can be instantly called on, used and developed during face-to-face exchanges.

When engaging in sensitive business environments consideration must be given to the interaction process. When attempting to persuade or influence others in a particular way, the first step is to determine the main objective – is it to achieve understanding, gain support or encourage others (forcefully or not) to take a particular course of action? When composing a communication strategy we should attempt to see how the other party may react to the message. Factors that might colour their point of view will include, for example, education, experience, background, culture, professional ethics and biases. It is also necessary to try to gauge what the chances are of achieving the desired result and the likely consequences.

Understanding others and ourselves is not always that easy. Luft (1984) developed an interesting perspective on things we know about others and ourselves. For every person there are things (wants, needs, likes, dislikes, goals, experiences, fears, etc.) that are open – things that we know about ourselves and that others can also recognise. There are also aspects which we keep hidden, things we know but do not share with others. There are also aspects that are blind – others see things which we do not recognise. Finally, there are some things that are unknown, things that neither we nor others can see (Figure 5.1).

When attempting to influence and persuade others we must always be aware that there are things that we do not know about the other person, indeed there will be aspects of the other person that neither the sender nor the receiver understands.

Social influence and persuasion

A few studies have examined the way individuals or groups use communication to gain compliance and influence others. A factor affecting social interaction is the power of each side to affect the other. Frost (1987) used the term 'surface power and politics' as a label for social influence that is used by an individual to get what they want from a decision, negotiation or interpersonal interaction. People who have, or

	Known to self	Not known to self
Known to others	*Open* (Common knowledge). Things that both others and we know about ourselves.	*Blind* Things that we cannot see, but others can. External behaviour observed by others but not recognised by ourselves.
Not known to others	*Hidden* (Personal secrets.) Things that we know about ourselves but do not disclose to others. What we don't want others to know.	*Unknown* Things that we and others do not know about. Could be discovered by self or others later.

Figure 5.1 Perceptions of ourself and others: Johari window. Source: Adapted from Luft 1984.

gain, greater power, use coercion, whereas those with less power tend to submit when a more powerful adversary uses power against them (Patchen 1993). Hare's (1976) review of communication studies suggested that regardless of whether a person's power is based on legitimacy, ability to co-ordinate group activity, skill, or some other factor, the more they attempt to influence others the more likely the success. This is especially true if the recipient is willing to accept the situation and peers do not set counter norms.

Persuasion and emotional interaction

Persuasion is the art of guiding, encouraging, convincing and directing others towards some form of preferred behaviour, attitude or belief. This is usually achieved through reasoning and/or emotional appeals. There are many different ways that we can communicate in order to exert a persuasive influence, discussed below; however, we must recognise that not all attempts at persuasion are effective. Some people may become resistant and resentful.

Emotion may be used to persuade people to adopt a certain point of view or take a particular course of action. Although reason and logic can be powerful tools, the use of emotion can be an important determinant of human behaviour. Hargie *et al.* (1999) identified a number of emotional methods of persuasion:

- *Fear and threats* When an individual is scared of another, or of a particular situation, it is likely that they will conform to instructions or threats as a way of dealing with their fear. Poor managers tend to use threats as a way of trying to control and influence those over whom they have some control; however, care is needed because apart from the obvious threat of alienation the workforce may well retaliate with threats of their own, e.g. the threat of strike action. Stress also needs to be considered.
- *Aversive stimulation* Subjects are exposed to unpleasant experiences, such as nagging, or a person loitering in close proximity. Such experiences constitute a mild form of physical or psychological torture. In the worst cases this constitutes harassment.
- *Consistency* This is the art of doing what we say. The consistency principle is very powerful. When we are seen to act on what we say, those working with us

tend to believe what we say – they can trust our commitment to something or believe that we will undertake action.

- *Commitment and ownership*   Further trust can be instilled by encouraging others to take ownership in decision-making processes. When parties are more involved in a decision they feel more committed to it. The combination of consistency and commitment can be powerful. Parties can be taken at their word, and where others are engaged they feel a sense of ownership and confidence in the process.
- *Morals*   Studies have shown that people can be encouraged to comply with a request if they are made to feel guilty. People can be reminded that they have a duty and/or a professional responsibility, it can be suggested that others may view their actions positively or negatively, or it could be suggested what the right sort of action is, what a knowledgeable, caring, professional person would do. The use of the moral argument can be a powerful tool.

The use of emotion during interaction should not be underestimated. Many of the reactions to emotion messages are conditioned responses that are developed from an early age. Although a logical and rational reaction to such interaction is often required, the part of the brain that deals with emotional signals is different to that which deals with logical information. When the emotional part of the brain is stimulated we may resort to deep-seated survival responses and reactions learned through previous experience. It can be difficult to deal with emotional interaction in a rational way. When it is realised that emotional tactics are being used in a discussion it is often useful to take a break, walk away from the discussion or ask for time to consider the information before taking any action. People react differently to emotional stimuli, thus when engaging in a meeting that might become stressful, it may be useful to be accompanied by a colleague to help balance reactions during such encounters.

Stress

Stress is a result of the way we internalise and respond to external occurrences. While the level of stress perceived is related to the actual events experienced, the way we view situations and react to them will also contribute to the amount of stress experienced. Some people seem to thrive on pressure, which would cause others to become ill. People deal with stress differently, and stress can develop during both positive and negative work experiences.

There is a link between stress and information processing. When people cannot understand information, or experience information overload they often experience stress. Stress manifests itself as an uncomfortable mental state. Sometimes we have difficulty in solving problems during stressful situations. We all have different thresholds and capabilities for dealing with information so when the information received is above our threshold or beyond our capabilities we suffer information overload. This can bring about uncertainty and stress, resulting in frustration and confusion. LeDoux's (1998) work on emotion and the brain provides a good insight into how information is processed during stressful situations. For example, when people start to experience heightened levels of stress they may be unable to process information with which they would normally be capable of dealing. During stressful situations the adrenal gland secretes a steroid hormone into the bloodstream. The release of the hormone helps the body mobilise energy resources so that we can deal with the stressful situation. Parts of the brain control the amount of hormone that is released. However, during particularly prolonged or very stressful encounters the brain fails to regulate the chemical release, and excessive levels are released into the blood. The chemical overload causes the brain to work differently and we may become unable to remember things, may experience difficulties in learning or simply struggle to make decisions. People who are exposed to prolonged or very stressful situations can suffer permanent brain damage. While some people work well under moderate levels of stress, others may be incapable of processing information during relatively similar experiences. Individuals and managers must seek to recognise stress thresholds and then manage workload accordingly. Treating everyone in the same way will lead to difficulties.

The use of emotion during interaction can result in stress. Threats and aggression can cause people to carry out the required task. Some managers may use such tactics to make employees work harder but this may result in a build-up of emotional tension in the recipient, which may not produce the results desired by the manager. High levels of negative emotion increase stress, and could possibly result in ill health or irrational behaviour. Care should be taken when attempting to use threats as the main way to influence others.

Disagreement

Disagreement is often seen as a negative term, yet it is found in most observations of group interaction (Bales 1950, 1970). Cline (1994) found that when groups avoid disagreement the vulnerability of a proposal may be overlooked, therefore a certain amount of challenge, evaluation and disagreement is necessary to appraise alternatives and reduce the risk of making a poor decision. Furthermore, Averill's (1993) review of anger-based research found that an angry outburst would often result in change that had positive benefits, and typically the relationship within which the anger was expressed was strengthened more often than it was weakened. However, people may choose to avoid disagreements to enable them to pursue relationship goals, believing that disagreeing would weaken the relationship (Wallace 1987, Cline 1994).

Argument versus aggression

A distinction has been made between argumentative and verbally aggressive behaviour (Anderson et al. 1999). Argumentativeness involves making refuting statements, whereas verbal aggressiveness involves attacking the self-concept of another. Although definitions vary, people seldom experience difficulty in recognising when we, or others, are aggressive (Averill 1993). Research has shown that group members who are argumentative express greater satisfaction with communication, and perceive their group as reaching higher levels of consensus and cohesion than do the members who are not argumentative but are verbally aggressive (Anderson et al. 1999). Argumentative members make contributions that are more rational and thorough than their less argumentative counterparts. Verbally aggressive members alienate other group members. Mild forms of aggression, such as threatening to break off talks, committing and sticking to one position, imposing time pressures on opponents or belittling the opponent's argument, are often used in negotiations (Pruitt et al. 1993), with varying degrees of success.

Agreement

Few studies have investigated the nature of agreement within groups and even fewer have considered what happens when group members seem to agree but in reality probably do not. During difficult tasks and stressful situations it has been claimed that members of the group were more inclined to pursue relationship goals that 'propped each other up' than to deal with the problem and enquire about the risks involved. Pressure to agree may be so strong that members continue to agree blandly while unwittingly consenting to their own destruction. Such attributes are associated with groupthink. Groupthink occurs when members of a group do not agree with statements that are made but do not make their view known to others, which results in the group members believing agreement is reached. Cline (1994) suggests that ways of avoiding groupthink include: asking questions, noting that an absence of disagreement should serve as a warning to group members to reassess alternatives, and knowing that the risk of illusory agreement appears to heighten as external stress increases. Hartley (1997) also points out that unanimous agreement may disguise the silent minority.

Opinion and beliefs

We have very different personalities with different ways of doing things, and this includes communicating. It follows that differences in personality can act as a barrier to effective communication. An individual's personality, beliefs, opinions and perspectives on life are said to make up their 'positive model of reality' (McCann 1993). In situations where we fail to take account of others' model of reality we are unlikely to be able to communicate effectively. It follows that we all should make an attempt to understand the viewpoint of those with whom we are trying to communicate.

Suggestive behaviour and disclosure of intent

People phrase things differently to alter the potential impact of the message. People will use different types of statement, e.g. present something as an opinion, suggestion, and statement of fact or proposal to generate different type of effects. Some people will make suggestions and proposals of how the group should act, while others may put the same point forward as their opinion or belief. The use of suggestions and statement of fact has a much harder impact on the group than someone offering an opinion. A suggestion identifies a course of action that should happen whereas an opinion identifies what a person believes to be correct. Some suggestions are autocratic and directive, they are given out almost as an order. The challenging of autonomous directions often results in conflict. Those receiving messages sent as opinions or even tentative proposals are often more comfortable challenging such statement than arguing against a hard suggestion. Indeed, some people phrase opinions so that that they encourage others to present their ideas. Those considered more effective in groups tend to use both opinions and suggestions; less effective members limit their interaction to opinions, beliefs and ideas (Gorse 2002). Group leaders and high status members have been found to use greater amounts of directive and suggestive interaction (Heinicke & Bales 1953).

Trust

Trust, communication and commitment are vital components in building a responsive and collaborative culture in construction. As interpersonal relationships develop over time and as reinforcing metacommunication is received the more likely an individual is to send out similar signals. Reinforcing interpersonal communication builds up an implicit knowledge of equity of effort towards communication. As each party shows their willingness to communicate the other will feel more secure in increasing communicative effort. The behaviour extends to tasks, which each undertakes as co-operative behaviour towards the other. Building of relationships is incremental and takes time. To trust someone there needs to be a certain amount of knowledge about the other's behaviour. We tend to trust information that comes from someone we have known for a while and believe to be a reliable source, i.e. we have had the chance to 'test them out' on a number of occasions. The realisation of expectations that we have of others' behaviour will often act as self-fulfilling prophecies (Wilmot 1980), i.e. when an individual acts in an expected way the prophecy is fulfilled. This can influence the way in which we behave towards others. This stereotypical behaviour may have positive and/or negative effects.

It is through the development and maintenance of interpersonal relationships, on an equitable basis, that interpersonal co-operation is developed. To send signals of a reinforcing nature there must be an inherent belief in what the other is saying. The receiver must trust the messages that he or she is receiving. Trust may be built up from previous experience of an individual, or from the knowledge that they are an expert in their field. Thus effective communication relies partially on the credibility of the sender (Arnold *et al.* 1996). Investigations carried out by Cook *et al.* (1979) found that some messages received from a communicator of low credibility would be accepted, although not immediately. They found the persuasiveness of the message would emerge weeks later, which is known as the sleeper effect, and where the source is forgotten but the message is retained.

There is an inherent need for trust within an interpersonal relationship. Research by Mellinger (1956) indicated that employees' communication behaviour was aggressive and evasive where there was low trust in their superiors. Smith *et al.* (1977) reviewed a number of studies that suggested trust was required for interpersonal communication to be effective, and that inconsistent or unclear messages made employees frustrated and anxious. Barriers within communication are evident whenever people meet. When communicating people trade on an equitable basis, it is uncommon to find a relationship in which one member of a dyad has disclosed more than the other member (Tubbs & Moss 1981). Reciprocal disclosure tends to be gradual, tending to take place only after a mutual trust has been established, with openness between both parties increasing as trust grows and more information is disclosed. The openness of the relationship tends to develop only after a basic level of trust and solidarity has been established. The polarised effect of the dyad can mean that individuals are careful of what is said, which may prolong the time taken for relationships to build.

Defensive and supportive communication

Defensiveness can be defined as the behaviour of an individual when they perceive threat from another individual or group. As individuals become defensive they expend energy towards protecting themselves. A person who perceives a threat may communicate in a guarded or attacking way. Defensive communication attempts to ensure that the information disclosed cannot be used against oneself. As people become defensive, behaviour patterns will be either consciously or subconsciously recognisable to other parties. The inner feelings of defensiveness create outwardly defensive postures. Where defensive communication develops without question, an increasingly circular destructive response occurs. The defensive signals distort the messages sent. The receiver attempts to understand the motives behind the defensiveness, rather than concentrating on the content of the message. As defensive behaviour continues it increasingly distorts.

Defensive behaviour makes communication less effective, whereas supportive or less defensive behaviour allows the receiver to concentrate more on the meaning of

Table 5.1   Defensive and supportive climates

Defensive climates	Supportive climates
Evaluation	Description
Control	Problem orientation
Strategy	Spontaneity
Neutrality	Empathy
Superiority	Equality
Certainty	Professionalism

Source: Adapted from Gibb 1961

the message. Gibb (1961) produced a list of interactive defensive and supportive climates that are said to help or hinder the communication process (Table 5.1).

If at any time, by any behaviour, the communicator expresses a defensive climate the receiver will be on their guard. The climates are interactive, so the defensiveness of an evaluative climate can be reduced by spontaneous action that shows support. The use of supportive communication techniques reduces the potential of defensive barriers to communication.

People have different models of reality and unless we are prepared to understand the perspectives of others we may not be able to communicate effectively with them. Good communicators develop a flexibility in interaction techniques enabling a greater appreciation of the needs of others (McCann 1993). Following this observation it is evident that we must be aware of all communication signals before we can respond in an appropriate manner. The content of conversation is only part of the message. Other signals such as tone of voice, tempo of speech, facial expressions and body language may change any literal meaning of the words involved. Failure to adopt a more sensory approach may lead to a partial understanding of the message. Equally, the notion of selective attention means that people are only able to take in so much information at a particular time. It follows that we must be aware of our own and others' strengths and limitations when engaging in communication.

Help-seeking and question-asking

Asking questions is the single most effective way to extract ideas and information although research has shown that where professionals do not understand a situation they may be reluctant to ask for help. Lee (1997) investigated the number of times a person sought help and asked questions and found that it is more likely that high status professionals will avoid situations where they need more information, in order to defend their status. Participants were less inclined to ask for help from higher status colleagues, and higher status colleagues were less inclined to seek assistance from others. In both cases advice was often sought in informal environments during conversations. Other research has also found that serious and costly errors have been made in multi-disciplinary projects, which could have been prevented by seeking expert help that was available. For example, Capers and Lipton's (1993) observation of engineers working on the development of the Hubble Space Telescope found that they avoided interaction with the specialists employed to provide expert optical advice, with disastrous consequences. Research on interaction during client briefing found that the construction specialists would rely on their own, limited, knowledge rather than suggesting that the contribution of other specialists would be useful (Gameson 1992). Professionals tend to avoid asking questions because help-seeking behaviour implies incompetence and dependence. Furthermore, most of us are not very good at asking questions (Ellis & Fisher 1994).

Help-seeking behaviour is fundamentally interpersonal, where one person seeks assistance from another. Seeking help often occurs simultaneously with information- and feedback-seeking from equal status peers and from those who have helped previously (Morrison 1993, Lee 1997). Co-operative patterns are reciprocal (Patchen 1993). Research by Gorse (2002) found that during site-based progress meetings the contractors' representatives considered to be most effective asked more questions than those considered to be less effective. Questions were often used to make others defend their proposals and to acquire more information, rather than to explicitly ask others for help.

Supportive climates and dysfunctional conflict

Supportive climates should not be seen as a way of avoiding conflict, but as a way of managing it. Conflict within communication can have positive aspects, reducing the risk of making a poor decision; however, if conflict results in a dispute, outcomes of a satisfactory nature are substantially reduced. The construction project should aim towards one common goal, that of the completed building. Functional conflict should help to solve problems; however, decision-making and problem-solving often lead to change.

Making informed decisions

Much of this chapter has addressed issues of persuasion, influence and argument. One of the problems faced by the project manager is how to engage with his or her multi-disciplinary group of professionals in order to make the best decisions. When working in a group, making an effective informed decision is about utilising the relevant specialist knowledge that exists within the group. Each person may have information that is relevant to the problem, some members will possess more knowledge on issues than others. What is important is that the most relevant information is accessed. Figure 5.2 shows a schematic of the ideal situation.

The most effective group decision will be made if those with the most relevant knowledge make a proportional contribution. It would be expected that those with the most knowledge contribute the most. Input from those with less knowledge which is nevertheless relevant, should still be considered: their different perspectives and experience may make those who are more knowledgeable alter their ideas. The combination of information helps inform the decision-making process. However, Diagram b in Figure 5.2 is less desirable, where those with less knowledge dominate the discussion and suppress their more knowledgeable counterparts. Those attempting to control the decision-making process should encourage contributions from specialists and invite balanced debate with those less specialised.

It is important to encourage some participation from all attending the meeting. There is always the exception to the rule and the best proposal may come from the person with least knowledge and experience. Good open debate, which allows opinions, proposals, challenges and disagreements should help ensure that all suggestions are properly considered.

Loosemore (1994) identified two factors associated with problem-solving in construction, which could lead to a defensive attitude. First, all problems involve a redistribution of resources (possibly meaning that some will benefit and some will not). Second, solutions to problems require something to change and the act of

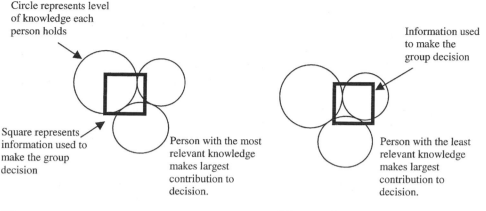

Good group decision

Circle represents level
of knowledge each
person holds

Square represents
information used to
make the group
decision

Person with the most
relevant knowledge
makes largest
contribution to
decision.

Diagram a

Bad group decision

Information used
to make the
group decision

Person with the least
relevant knowledge
makes largest
contribution to
decision.

Diagram b

Figure 5.2   Effective and ineffective use of group knowledge.

change is not attractive to many people. So, dysfunctional conflict may emerge in construction projects as organisations defend their allocation of resources. It is inevitable that things change and evolve and it is impossible to predict all eventualities. In large organisations and projects the main function of senior management is the adjudication of competing requests and conflicting demands (Katz & Kahn 1967). Failure to deal with the situation quickly and effectively will lead to conflict.

Conflict

During interpersonal communication it is possible that differences of opinion may emerge, and where these are firmly held beliefs the result will be some form of conflict. A certain amount of conflict within any organisation is inevitable; however, the existence of communication problems will make the management of conflict difficult. This is particularly true of construction projects with their temporary and fragmented arrangements. Organisational conflict may occur between two people or within a group or team. Due to the short-term nature of the temporary construction project the participants are developing relationships for the first time, while also seeking essential information to do their particular job. Problems may be experienced during the formative stages of the relationship, due to parties holding back information until trust is established. Although information needs to flow without obstruction, the reciprocal effect of interpersonal relationships may cause problems during the early stages of the relationship.

Individuals have a unique personal history having been exposed to different situations, environments and behaviours, and develop different perceptions. This goes back to selective attention as individual memories of past experience influence what they see as relevant in current situations. Perceptions can be seen as the processing and development of recognition and identification with our environment. However, where two people are in the same area, in the same physical environment, not all of the information within that environment will be processed, and different information will be processed in different ways, at different times by the two individuals. Different people will have different experiences of the same situation. The phenomenon of individuals selecting and processing different stimuli to different extents has been supported by research (Huseman 1977).

Ambiguous communication can often lead to confusion and hence conflict. In complex and uncertain situations it is imperative that the message is clear and unambiguous. Complexity and uncertainty can often result in an ambiguous message which, when disseminated, further complicates the issue, leading to misrepresentation and distortion. If roles and responsibilities are unclear, construction professionals may interpret their responsibility in a way which suits them or remain confused over what it is they are supposed to do.

Conflict can be viewed as a positive aspect leading to development, innovation and flair, or as a detrimental and destructive process, i.e. it can be functional or dysfunctional. Natural conflict is described as the intended or actual consequence of encounter resulting in stronger participants benefiting from the clash. It is further suggested that good conflict is goal orientated rather than disagreement over the goal, i.e. 'What is the best way of achieving?' rather than, 'What are the individuals attempting to achieve?' In situations where the actions of one person do not affect others, competition is good, possibly producing better results than co-operation; however, where participants are interdependent, co-operation is often more effective.

Further reading

Luft, J. (1984) *Group Process: An Introduction to Group Dynamics*, Mayfield, Palo Alto, CA.

6 Group communication

Group behaviour and communication are central to the design and construction of buildings. In this chapter we explore the dynamics of the small group and group decision-making. We then look at some of the essential factors required to enable groups to function, covering participation in groups, and the issue of reluctant communicators and group development. From this we look at intergroup communication and the management challenge.

Group development and group norms

The importance of group and team development is well documented in management literature and new developments in teamwork and groupwork strategies have become an important theme in management literature (e.g. Drucker 1995, Hartley 1997). Construction relies heavily on the co-ordination of many different specialists and this is usually referred to as a team effort, with the project being the focus and hence the *raison d'être* for the team. We have taken an alternative view here, our premise being that construction projects are realised by the co-ordination of specialist groups, teams and individuals that collectively contribute to the project. Communication between the management and design during the construction process is, for the most part, a function of group interaction. Individuals work within small groups in their own organisation and with other, complementary, groups in different organisations, combining their skills and knowledge to achieve the project outcomes through co-ordinated activities. To do this effectively each individual must work within a structured and organised group. The contribution of more than one individual to a problem increases the number of perspectives, the depth of expertise and knowledge and the amount of information available from which to make informed decisions. However, the effectiveness of the group and the degree of co-operation between members can depend on the communication strategies employed and the training provided. Dimbleby and Burton (1992) simply state that group communication occurs within groups of people and by groups of people to others. Other researchers suggest that individuals forming the group need to share common attributes, goals and/or interest (or at least have common values of norms of behaviour) for communication to be effective. For Kreps (1989), small group communication occurs among three or more people interacting in an attempt to adapt to their environment and achieve commonly recognised goals.

Construction projects are multi-disciplinary, in that they bring together professionals with different specialist knowledge from different organisations to ensure the various aspects of a project are achieved within the project parameters. This temporary social system comprises small groups that collaborate on, and contribute to, tasks to achieve a common goal through the use of multi-group communication. Various communication practices will be used in order to realise the building, and issues surrounding the group communication and the decision-making process must be considered. Productive groups have been found to have a structure that is suited to their function. The high level of productivity is achieved not only because they have procedures for solving problems, but because the group is stable and less time is devoted to status struggles (Heinicke & Bales 1953, Hare 1976).

Group development

Since the achievement of a group's goals depends on concerted action, members must reach consensus on acceptable task and socio-emotional behaviour before they can act together (Hare 1976). The social element of interaction is developed through emotional exchanges that are used to express a level of commitment to other members. The level of interaction associated with maintaining and threatening relationships (socio-emotional interaction) will be subject to group norms. Task interaction is related to the exchanges of opinions, information and suggestions that help facilitate group activities and goals. To achieve group goals individuals must engage in socio-emotional interaction to maintain the relationships and task interaction to co-ordinate group activities.

A group's behaviour will develop and change over the period of interaction. As task groups attempt to solve problems, moving towards a solution, they undergo changes in terms of their attitude and behaviour towards each other. Groups go through a process of learning, which can result in changes to their structure as the group moves through a range of social, emotional and developmental stages. Two variables that affect group development are the length of time that a group has existed and the number of occasions that the group has met. Borgatta and Bales (1953) found that when people have taken part in a series of meetings on related subjects and different people are present in each of the previous meetings, group participation is the same as if the group had met for the first time. Bales' (1950) early work found that this phenomenon is due to the group's socio-emotional development, i.e. individuals not being aware of the group's social and emotional norms, and the group not knowing how the individual will react to the norms. Thus, a socio-emotional framework develops and re-establishes itself when new individuals enter the group. There are parallels here with the site progress meeting at which different participants attend over the course of the contract.

Development of group norms

Although the behaviour and characteristics of groups change and develop over time as the group adapts to its environment, it is also well known that groups develop and are subject to behavioural norms. Newly formed groups, in the course of time, tend to develop relatively stable patterns of interaction leading to familiar patterns of behaviour (Keyton 1999). Anderson *et al.* (1999) make a distinction between rules and norms, noting that members come to accept norms as their way of being a group and doing group work, whereas rules are agreements about how to behave appropriately. The norms of group behaviour may be specifically associated with the reason why the group was formed, or they may be attributable to the group make-up.

In most situations there are a number of specified roles or repertoire of acts that provide information about how the individual is supposed to interact, and these vary from one situation to another. Expectations of the way group members are supposed to act are articulated into implicit rules that are adopted by the group to regulate its members' behaviour (Fledman 1984). Such norms and rules are said to provide powerful controls over the group. While there are rules and norms which are explicit, it is those that are implicit that have the greatest direct effect on relational behaviour (Keyton 1999). Norms are often the least visible yet most powerful form of social control that can be exerted on a group.

Group norms can be so influential that some individuals will express a judgement differing from the one they hold privately (Hare 1976, Hackman 1992, Shultz 1999). Fledman (1984) has identified four ways in which norms are developed:

- From statements made by leaders
- Critical events in the group's history can establish a precedent. For example, when members are faced with a deadline the group may change its pace of interaction to ensure the deadline is met
- Simply develop from repetitive behaviour patterns. Such patterns are particularly prominent in certain seating configurations
- Members can import group norms from previous group experiences.

It is essential for newcomers to observe the communication behaviours and practices of other members, so that they can understand the group culture and participate in it; this is a period of socialisation and acceptance. When new groups form they establish beliefs, values, norms, roles and assumptions that are specific to the group. An individual's actions and behaviours are also influenced by his or her motives for membership, positions and role (Zahrly & Tosi 1989).

Norms and decision-making

A number of case studies focus on how norms are used by groups (e.g. Hirokawa & Salazar 1999). These and other studies have shown that rules and norms are habitual, forming a backdrop or structure against which decisions are made and can have both positive and negative effects on the decision-making process (Janis 1982, Larson & LaFasto 1989, Hackman 1992). They can encourage cohesion and agreement, suppress critical enquiry, reduce political input and increase rational discussion. Giles (1986), looking more specifically at language, noted that communication behaviour reflects the norms of the situation. However, it is often the communication behaviour and language that are used to define and subsequently redefine the nature of the situation for the participants involved, from which decisions are made.

Deviators

It would seem that group norms affect all members of the group, however, Keyton (1999) has suggested that high status members may be exempt from norm expectations. If a member deviates from the group norms other members tend to react in one of three ways (Hackman 1992):

- Group members may try to correct the behaviour, normally through pressure outside the group environment
- If deviation persists, other group members may exert psychological pressure through communication, placing the deviant in an 'out-group' position
- Finally, if deviation presents an acceptable alternative to the group norm and continues to maintain this stance, over time it can influence other members to accommodate the alternative norm.

Equilibrium theory of group interaction

Bales and Strodtbeck (1951) found that problem-solving groups exhibited recurrent patterns of interaction, which they identified as orientation, evaluation and control.

- *Orientation* The initial phase 'orientation' is marked by high levels of task-related messages in the form of information, opinions and suggestions, and positive and negative reactions to this information. This phase involves communication about the nature of the problem to be solved.
- *Evaluation* The second phase is characterised by a reduction in exchanges of informational acts, a levelling out of opinion and evaluation, accompanied by suggestive behaviour and positive and negative reactions. During this phase the group confronts 'what to do and how to do' type issues.
- *Control* In the final stage, the group establishes 'control', this is marked by a continued and sharp decline in informational behaviour, a slight decline in opinionation, a reduction in the quantity of suggestions and negative emotional behaviour, and continual increase in positive reactions.

The control phase involves deciding what to do. At the same time as the task-related acts are discussed, a parallel cycle of positive socio-emotional phases results; interaction acts such as showing solidarity and tension reduction are used. In order to address problems, groups have to move through this set of acts. To ensure that task-based discussions can continue, the relationships are maintained with positive and negative emotional exchanges.

The work of Bales on group interaction and development was followed up by a number of researchers. The most popular theory, cited in many management books, is Tuckman's model of group development. Using observation methods adapted from Bales (1950), Tuckman (1965) suggested that there were four stages of group development, being:

(1) *Forming* During the early stages of group development, members tentatively get to know each other; they are polite and careful not to cause conflict. Individuals are primarily concerned with being accepted into the group. Group behaviour during this stage is inhibited as members give and receive information.
(2) *Storming* As individuals start to feel more secure and accepted by the group they start to put forward their own ideas and opinions more forcefully. As the group matures, members confront their differences and a level of conflict emerges. During this period group members will start to learn where and when conflict is likely to occur, and what issues will cause disputes.
(3) *Norming* If the group survives the storming stage a group framework will develop. Explicit and implicit consensus will be reached on roles, power, status and procedures. Agreement on such issues results in a reduction of hostility and conflict. During this stage groups become more cohesive.
(4) *Performing* The group's norms, which provide the accepted processes and decision-making structure of the group, help the group perform better. Little conflict is experienced during the performing stage.

Tuckman and others have since added a later stage to the model. The phase known as adjournment, or mourning, recognises that towards the end of a group's life cycle the group's activities subside, members start to leave, and levels of activity drop. Following the experience of very successful groups the members may experience a sense of loss, believing that they may never work in such a successful group again. As a result of such feelings, their motivation may temporarily drop. In organisations some groups may start to disband and reduce their activities before projects are complete. Considering the number of phases a group has to go through to reach the performing stage, team leaders should be careful to maintain group activity, ensuring the project is complete before the group breaks up.

Theories such as the equilibrium theory, because of their single-dimension approach, became known as linear-type theories. Scheidel and Crowell (1966) were the first to question this theory: they noted that interaction did not unfold in a linear way. The interaction is determined by a set of internal relationships that emerge from the elements composing and characterising a group; at the same time the group is influenced by other elements from external environments. Poole (1981) has also challenged the theory of linear interaction development and also found evidence to support the multiple sequence models. Their view was that group interaction progressed in many different fashions, rather than through linear development. Group acts are contingent on task and situational factors, and multiple sequences of phase movement are caused by such variables as task characteristics, group composition and the level of conflict evoked by task issues. Bales (1970) also claims that the same factors affected interaction, although this seems to have been overlooked in his earlier work. The arguments over single and multi-dimensional methods of studying interaction continue; however, theories developed from both approaches have contributed to the understanding of group dynamics. Neither single nor multi-dimensional approaches offer a comprehensive picture of the nature of group communication, but both provide valuable insights into the interaction process.

Group participation and interaction

When groups or individuals meet for the first time, they arrive with certain assumptions about the roles of the various participants, which will vary according to why the group has formed (Bentley 1994). People develop fantasies about others, usually based on their assessment of physical features, profession, behaviour and nationality, etc. The behaviour and interaction of professionals on past projects will affect the way they communicate on new ones, simply because we expect people to use the same (or very similar) communication behaviours to those previously used. This provides a (false?) sense of security because we also expect the communication strategies to work, despite the fact that some groups will have different members and the composition of the project groups will differ from those previously known to us. We must accept that issues of developing group norms and intergroup communication need to be addressed early in the project to establish clear and efficient communication routes. Failure to do so may lead to ineffective communication and early communication breakdown within and between groups.

Participation is the extent to which individuals are involved in group interaction and is coloured by the group norms and group development. It involves aspects of turn-taking, initiating conversation, interrupting and the intensity of interaction (Ketrow 1999). Littlepage and Silbiger's (1992) research on participation and turn-taking have found that interaction is controlled and dominated by a few members of the group. In moderate and larger sized groups, it is widely accepted that participation among group members is skewed and unevenly distributed. Bell's (2001) study of multi-disciplinary teams concluded that the limited contribution of some specialists prevented a truly multi-disciplinary perspective being presented, but the skewed participation in groups does not always hinder performance. Although participation is uneven there is evidence to suggest that group members become more dominant when issues associated with their particular specialism become more important. Wallace (1987) found that different communication tactics were used in order to control specialist contributions. His observations of construction design team interaction found that participation is a function of the group's characteristics, with participation varying in relation to the way the group develops. An individual's participation in the group's interaction is regulated by the feedback or response received in relation to the previous contribution made. Types of feedback or response signal include turn-taking signals, attempt-suppressing signals and back-channel signals. Turn-taking and suppressing signals are given by the current speaker; they are used to defend the right to continue speaking on the same subject or with the same level of emphasis. Back-channel signals are communication acts by others, such as a person agreeing or disagreeing with the speaker. The types of signal and the rate at which they are used relate to the underlying group process, particularly the group regulatory forces. Meyers and Brashers (1999) found that groups use a form of participation reward system; those who are co-operating with the group receive helping communication behaviours and those in competition are received with communication-blocking behaviour.

Reluctant communicators

Participation in a group is also related to an individual's willingness to speak. This may lie outside the direct influence of the group process and development (Wallace 1987); however, an individual's reluctance to communicate may affect the group's participation process. Burke (1974) suggests that our willingness to communicate accounts for most of the participation during group interaction, assuming that communication takes place in a democratic group environment. People who tend to avoid communication are termed reluctant communicators (Wadleigh 1997). McCroskey (1997a, b) found that shyness may occur due to communication discomfort, fear, inhibition and awkwardness. Some people will initiate communication while others, under virtually identical situations, will not: these latter are the reluctant communicators. In group situations apprehensive individuals talk less, avoid conflict, and tend to be perceived more negatively than members who talk more. Highly apprehensive people also have a tendency to attend fewer meetings, although this reluctance tends to diminish over time.

Reluctant communicators are unlikely to hold influential positions or be seen as leaders by their peers. Relationships have been found between perceived leaders and high levels of verbal participation. For example, Mullen *et al.* (1989) found that the individual perceived to be a leader by group members and observers was the most frequent contributor, being responsible for 50 to 70 per cent of the participation. Bales (1970) found that talkative group members attracted attention to themselves through their domination of group interaction, and although this may result in other members attributing leadership to the most talkative member, the leader and director of the group is usually one of the quieter (but more persuasive) participants. As well as reluctant communicators, there are individuals who interact more frequently than others. In decision-making groups, those who talk the most tend to 'win' the most decisions and become leaders, unless their participation is excessive and thus antagonises the other members.

Maintaining relationships

The primary issue facing work-orientated groups is the need to maintain a balance between task and social demands (Keyton 2000). Bales (1970) found that as groups address problems emotions start to develop and, as a result of disagreement, tension is built up between members as they focus on the problem rather than relationships. Bales' observations noted that conflict, even when constructive, leads to tension that can damage the cohesiveness of the group and threaten group maintenance; however, too much attention to cohesion stifles constructive conflict and

threatens the group's ability to solve problems. Cline (1994) identified the importance of functional conflict to avoid 'groupthink' and improve the decision-making
process, although conflict may also damage relationships between group members.
Conflict often emerges from perceived failure, thus moderate levels of conflict are
needed to avoid failure but at the same time increase productivity. Tension
resulting out of conflict may be removed by positive emotional acts (such as joking,
and praise) and negative emotional acts (such as disagreements, expression of
frustration and even aggression). If socio-emotional issues are not addressed when
they arise, the increase in tension may inhibit the group's ability to progress in its
work. Groups must maintain their equilibrium, moving backwards and forwards
between task and socio-emotional-related issues.

Too much attention to task interaction can limit the communication required to
build and maintain relationships. If groups are to perform effectively, positive
reinforcement, including agreeing, showing solidarity, being friendly and helping
release tension, are needed to offset negative reactions. Bales (1953) found that
when group members had dealt with a problematic task they would diffuse
negative emotions with positive emotional discourse, returning to the task issues
once the tension had been dissipated. Group members prefer positive feedback and
interaction which suggests that the group is effective, which increases morale.
Gorse (2002) found that contractors who were considered more effective used more
negative and positive emotional exchanges than less effective contractors and the
level of positive emotional acts was greater than the negative exchanges. While
significant, the level of positive interaction was not as high as previous studies
suggested was necessary, being just 1 to 4 per cent greater.

The distinction between task and socio-emotional behaviours remains a fundamental assumption of group communication research (Poole 1999). There has been
a tendency for scholars to believe that task and social dimensions are in competition
with each other, with the result that many studies have a greater emphasis on the
task-based factors, neglecting social relationship issues (Frey 1999, Keyton 1999).
Keyton (1999) points out that even when research does consider emotions, they are
usually considered with respect to their impact on task messages or outcomes
rather than their impact on relationships. Relational acts are often found to facilitate
the group development process but inhibit group performance.

Leadership: task and relationship roles within groups

Group members must undertake roles to ensure that task and maintenance goals
are maintained. Members judged to hold positions of leadership have been found to
have certain tendencies. The most frequent talker tended to be the most highly
respected but most disliked member of the group, a role referred to as the task
leader. The next most frequent talker was not as respected as the first but tended to
be the most liked member of the group, the maintenance leader. A study of interaction leadership traits by Heinicke and Bales (1953) found that individuals who
held high status would participate and contribute the most during early meetings,
with their contribution reducing in subsequent meetings. At first Bales (1950) failed
to examine the extent that leaders were associated with task and maintenance
functions. The second problem that emerged was that the participants' ratings
implied a distinction between two different types of task leader (Pavitt 1999):

- The best ideas person, also known as the substantive leader
- The procedural leader the person giving the most guidance.

While the split between different types of leader in a group has been questioned, the functions of leaders have largely been substantiated. Pavitt (1999) noted that the distinction between task and maintenance leadership functions, as well as the further divisions of the task function into substantive and procedural, appears to be sound. However, the distinction drawn between task and maintenance functions has received criticism. Wyatt (1993) found that in certain situations, such as therapy and support groups, the task is to build relationships, taskwork being the same as maintenance, and a communication act could actually serve both roles. The extreme role differentiation between task and maintenance leader sometimes appears artificial, as such roles often change between meetings.

Early studies of group behaviour using Bales' (1950) method found that groups exhibited regular patterns of interaction that were specific to the context in which they were observed. When the context of the group setting was highly controlled, small changes in the group size did not have a profound effect on the behaviour of the group. However, subtle differences in socio-emotional behaviour have been found to produce significant changes in the group's behaviour. In Wallace's (1987) study of the construction design team, social and emotional interactions were used by the group to nominate and elect the group leader and support their status. As the group tasks changed, socio-emotional interaction was used to remove those elected, enabling others to become more influential.

Multi-disciplinary groups

Despite the large amount of literature on group development and communication, the vast majority of it has been based on devised experiments. Very few of the studies are based on real-life groups trying to go about their business in the workplace. Bell's (2001) work on multi-disciplinary team discussions found that high levels of task interaction, ranging from 83 to 93 per cent, typified the discussions. When the proportion of giving information, opinions and suggestions was compared with asking for information, opinions and suggestions, all of the professionals except one gave, rather than asked for, information. Such observations are consistent with studies by Gameson (1992) and Gorse (2002).

Bell's findings were consistent with previous studies of groups and influential members. She found that the interaction within the groups was not evenly distributed. Where an agency involved in the meeting was represented by more than one member the senior representatives would make a greater contribution than the less senior representatives. Interaction was not evenly distributed across the group and would be dominated by one or two members. As the group sizes increased the proportion of members contributing to the meeting decreased; this is consistent with reports by Bales. Bell concluded that the lack of contributions made by many of the specialists meant that the multi-disciplinary teams failed to provide a holistic view. This observation raises a question as to whether specialist perspectives are fully utilised in multi-disciplinary team meetings.

Group performance and outcomes

Clampitt and Downs (1983, 1993) note that intuitive links between communication and productivity make sense. They also cited a number of surveys that showed perceptions on this relationship were strong. However, they found that perceptions of productivity were diverse and the link between communication and performance was considerably more complex than had previously been assumed. Valid criteria for judging the effectiveness of real-world decisions are difficult to define and may conflict. What might appear to be a successful short-term decision may result in long-term problems, and vice versa. Poole *et al.* (1999) found that when different groups of people evaluate group performance, differences are often found. External evaluators have been found to evaluate group performance differently, taking a more negative view of group performance than group members.

Group decision-making

Group decisions will be affected by the communication media used and the cultural setting of the group, such as group membership and organisational identity. External conditions (such as workload, time and pressure) and internal conditions (stress and emotion) can affect the ability of group members to work effectively, colouring and changing their patterns of communication within the group. Participation within the group depends on the social norms of the group, which will differ across various cultural settings.

When individuals work in groups to solve problems they need to use their individual knowledge to inform the group decision-making process. The group must access and discuss all relevant knowledge possessed by the group and use their combined skills to evaluate the information and arrive at the best decision. However, decision-making is complicated and the time allowed for decision-making is a determining parameter, as is the amount of information available to the decision-makers at that particular point in time. We must attempt to make maximum use of the resources available in order to make a decision within the allocated period. A snap judgement based on incomplete knowledge or personal instinct is risky, and may often be inaccurate and unfair. However, we are often faced with situations that require an immediate response and so there may be little time to consult information. Even with extra time to solve a problem, a decision made based on the information available on one day may be different from that which would be made on the next when more information is available. So, unfortunately, there will be occasions when we have to rely on our intuition, perhaps because of insufficient information or lack of time to consider the options. On our own this can be risky, but in a group situation there are other members to question and/or support the intuitive decision, thus giving some reassurance. This is particularly true of construction projects. With pressure on reducing costs has come pressure on time (which is an expensive resource) and so the majority of those working in the construction sector are trying to do a lot in a very tight programme. This applies to designers and to contractors equally. One negative side effect is that the production information is rushed and is often delivered to site incomplete and containing errors. This places additional pressures on the site personnel who not only have to spot the discrepancies but also have to request additional information from the designers. Obviously there is considerable pressure to ensure that the work is not disrupted and so it is often necessary to make a quick decision so that work can proceed. However, given the implications of getting it wrong on a construction site we would urge all 'pressured' decisions to be discussed with a colleague before issuing the appropriate instructions. Whether the decision turns out to be a good one will only become evident with the passage of time.

Time constrained discussions and 'closure'

Problem-solving during the construction process is subject to time pressures, and these problems need to be resolved or 'closed' if the programme is to be maintained. De Grada *et al.* (1999) found that the time pressure prevented groups from engaging in 'social niceties' hence resulting in groups emitting a lower proportion of positive socio-emotional acts. The time constraint also encouraged a conversational pattern wherein some members manifest greater dominance of the discourse than others. De Granda *et al.* found that those with a greater tendency towards resolving problems quickly were more dominant in time-constrained discussions. They also tended to adopt more autocratic styles of leadership, giving directions to other members of the group. In commercial projects that are constrained by time factors the professionals make many decisions, normally over a series of meetings, and they have to sustain relationships sufficiently so that they can work together during their involvement in the project.

Conflict is an essential part of organisational activity. Research has found that arguments and disagreements are sometimes avoided so that members can pursue relationship goals. The balance between disagreement and agreement in a project environment may be difficult. During group meetings, issues have to be discussed with sufficient rigour to produce the optimal solution, but relationships must be sufficiently maintained so that members are able to continue to operate effectively. Failure of participants to make effective contributions to group discussions will reduce the group's decision-making ability. Some members of the construction team may be reluctant communicators. Others, while active communicators, may restrict their interaction to task-based discussions, avoiding emotional exchanges. The temporary nature of the construction team may also restrict the nature of interaction. During early stages of group interaction, the participants confine communication to task-based messages; emotional exchanges emerge as members become more familiar with each other. Where specialist knowledge is required and the professional does not make a full contribution, the ability of the group to make an informed decision is reduced.

Risks in group decision-making

Literature on group performance and multi-disciplinary working suggests that the decisions made by groups are more workable, more accurate and more rational than those made by an individual because of the wider range of knowledge available to the group. Stroop (1932) argues that the grouping of knowledge and experience acts as a moderating influence to restrict extreme views. The group's regulatory forces, which are imposed using conflict and group norms, control unacceptable views that are presented to the group. Contrary to this view, work by Rim (1966) on group and individual risk-taking found that group decisions were more risky than those of individuals. Rim found that 13 of the groups adopted higher risk strategies to problems following group discussions. Thus, group interaction may change the behaviour of individuals. Bemm *et al.* (1970) found that individuals within groups would take greater risks even if the consequences of the risk-taking would affect them personally. However, where group members were informed that failures associated with risk-taking would be openly disclosed to the group, there was a shift to less risky decision-making. Although such findings have been compared to the 'real world' context, the findings remain limited to laboratory-type experiments. Due to the complexity of commercial problems it can be difficult to identify the level of risk involved in a decision, and whether the distinction made between individuals and groups applies to commercial decisions.

Brainstorming and idea generation

Early research by Stroop (1932) found that group interaction produced a higher degree of creativity in relation to the solution of a problem than an individual. Others subsequently noted exceptions to this observation. Research on idea generation through brainstorming exercises has shown that individuals outperform the group by a factor of 2:1, and the individuals' ideas were found to be more creative than the group's ideas. The main finding from this research was that group pressures inhibit members' participation. Individuals participate less in small groups when they feel that their skills are inferior to those of others (Collaros & Anderson 1969). Individuals tend to contribute more ideas when working in isolation; however, evaluation of those ideas may be better dealt with in groups where different perspectives can be used to analyse ideas.

Deciphering relevant information

A prime obstacle to solving 'real-life' problems is selecting the relevant data from the body of superfluous, irrelevant and possibly misleading data. Experimental research by Campbell (1968) found that subjects take longer to solve problems where they had to differentiate between irrelevant and relevant data. Furthermore, the time required to solve the problem increased as more people became involved in the problem-solving exercise. In commercial environments participants must explore the options and solutions available when solving problems. Given the vast amount of information available to all members of the construction project the issue of relevance needs to be addressed, as does an appreciation of the user's requirements.

Multi-disciplinary and uni-disciplinary groups

In uni-disciplinary groups the objectives of each individual are likely to be similar to those of other members, while in multi-disciplinary groups there is likely to be larger variation in objectives (Wallace 1987). Multi-disciplinary teams have been found to propose and consider a wider range of solutions to a problem when attempting to arrive at an overall solution (Ysseldyke *et al.* 1982). Although Bales suggests that multi-disciplinary teams may appear more productive in terms of alternative solutions generated during interaction, this could also be a result of goal ambiguity. Yoshida *et al.* (1978) examined the content of multi-disciplinary group interaction. They classified the interaction into five main categories: contributing information, processing information, proposing alternatives, evaluating alternatives and finalising decisions. This study found that the frequency of the individuals' participation and their perceptions and contribution to multi-disciplinary teams varied more than uni-disciplinary teams.

Recognising expertise

The findings of Yoshida *et al.* identified that the stronger combined group forces often overruled individual expertise and experience. Thus, group consensus may go against expert opinion and information. However, work by Littlepage and Silbiger (1992) found that, regardless of uneven and skewed participation rates, groups were able to recognise and use individual expertise confidently.

Overt communication

Effective communication in decision-making is required to transfer understanding of the problems and hence discuss the various options that are available to the group. Hosking and Haslam's (1997) observations of business relationships found that informal conversations within organisations were an important process for understanding what were considered as 'taken-for-granted' statements and to help group members to overcome ambiguity. Hollingshead (1998) found that when members of a group are tasked with a problem, members become specialists in some areas but not others, and all members come to expect each participant to access information in specific domains. The specialisation reduces the cognitive load on the individual, while providing the group access to larger amounts of information. Those in new relationships or groups must communicate to identify explicit responsibility for gathering and processing specific information. Assumptions about responsibility for problem-solving result in less effective teamwork and duplication of tasks undertaken.

Broad use of communication techniques

Effective groups in industrial settings are those that are more productive and meet the organisation's objective. Shepherd (1964) suggests that successful groups have open and full communication in which information, ideas and feelings are exchanged and no one holds back. Gorse (2002) found that construction groups and individuals that were considered more effective used a more distributed range of communication techniques. Individuals considered to be most effective showed more positive and negative emotions, asked more questions and gave more direction. Those considered less effective limited their interaction to information- and opinion-giving, there was little use of question-asking and they hardly entered into emotional exchanges.

Intergroup communication

The way a group develops interaction norms will affect the group's ability successfully to discuss tasks, evaluate proposals and maintain relationships. The findings of research suggest that question-asking, giving directions, and use of emotional interaction can affect group behaviour. The development of communication mechanisms that either inhibit or enhance the exchange of information within the group is fundamental to the collective performance of the project. Research seems to suggest that groups require direction, yet equally it is important that members contribute using a broad range of communication techniques to develop proposals and make decisions. From the perspective of the construction project an important factor relates to how well groups work together, i.e. how they communicate. This is taken up in Chapter 8 under organisational communication; however, it is worth noting that the way groups work is a fundamental prerequisite for understanding and then designing the communication networks (see Chapter 7). Consideration needs to be given to:

- Recognising group dynamics
- Building and maintaining groups
- Facilitating intergroup and interorganisational communication
- Encouraging and incorporating feedback from groups.

The management challenge

All managers must make an effort to understand the dynamics of the groups that they are tasked with managing. Managers must try to facilitate the group's efforts by designing managerial structures that provide an adequate framework, but which are flexible enough to allow the group to achieve its tasks in a creative and purposeful manner. Key to this is the manner in which different groups communicate – this cannot be left to chance. The communication networks must be put in place that encourage and facilitate open communication, thus allowing for the transfer and incorporation of expert knowledge. This is a particular area of concern for the design manager and the construction manager. It is also related to the design of the project culture and so is a prime concern of the project manager (see Chapter 9).

Further reading

Hartley, P. (1997) *Group Communication*, Routledge, London.

7 Communication networks

Research has shown that identifying networks is a difficult task, but essential if organisational communication is to be understood and managed effectively. The management of formal and informal communication links is essential to control the distribution, interpretation and the effects of information exchanged. Diffusion of information and knowledge around networks is discussed, as is the role of organisational gatekeepers. We conclude with some thoughts on supply chain management.

Unstable networks

It is common practice to use network analysis techniques to determine the communication structure within a social system, from which communication networks may be represented graphically by a sociogram. Analysis of the frequency of communication between individuals within a social system can identify the most active lines of communication and gatekeepers, etc. However, there is a methodological problem with sociograms, since they represent a network at a fixed point in time – they do not address the change in the network over time (Rogers & Kincaid 1981). Furthermore, there is a problem with the data collection in knowing exactly when the communication link reported by the respondent in response to a sociometric questionnaire actually occurred. Thus according to Rogers and Kincaid the dynamic process of communication relationships is 'so fleeting that networks cannot be accurately charted' (Rogers & Kincaid 1981: 314). Indeed, researchers trying to map information flow in construction projects have, by their own admission, not been able to do it; rather they have helped to highlight the magnitude of the problem facing researchers. There is a dilemma here. If communication networks cannot be charted accurately, then how is it possible to manage the (temporary) communication networks that develop for individual building projects? The answer, of course, is that we have to make assumptions and try to model the communication networks to the best of our knowledge. In construction there are other difficulties. The communication networks are specific to individual projects and therefore temporary in nature. Networks are not stable, they change as the project develops with different individuals and organisations entering or leaving the network coincident with different project stages.

Each project will have an associated group development phase where individuals enter and leave the project at different junctures, thus the composition of the project network will be flexing and evolving throughout the life of the project. So rather than one project-specific network it may be more realistic to see the project as a series of overlapping networks. This makes it particularly difficult to manage – but its effective management and direction is vital to the effective realisation of the finished building. Individuals allied to the project may be linked by communication channels or information exchange routes, which will be both formal and informal. The information being transferred around the various networks will be of two types, either project specific (process information) or product specific (product information). The majority of this information is task specific and is unlikely to be shared by anyone other than those who need to use the information. All this conspires to make the project manager's job a challenging one.

Information management

The generation, transmission and filtering of information as it passes around the network is a particularly complex area. Essentially, each project could be viewed as a problem, a problem broken down into a series of smaller problems to be dealt with by different individuals. In order to solve the problem and communicate the solution to those who need it, information has to be gathered, analysed, filtered and assembled in a new arrangement. This information is then transmitted on to the next link in the chain, which may or may not need all of the information they receive; indeed, they may need additional information that the sender has failed to send. The quality, rather than the quantity, of that information will ultimately determine the quality of the finished building and the service provided by the organisations working on the project. The quality of the links between individuals, otherwise known as the nodes in the information network, is essential to effective communication.

Effective communication and information management is vital not only to the efficient day-to-day running of organisations, but also to the daily administration of individual projects. Information needs to be managed so that parties receive all the relevant information without being overloaded and without having to gain the information through informal channels. People cannot deal with information that exceeds their processing ability; information must be transferred so that those with skills, knowledge and capabilities can use the data. Some information processing tasks in a construction project are simple and can be dealt with by one person; however, many of the problems are multifaceted requiring the contribution of various specialists. Multi-disciplinary problems may be better approached in a sequential fashion with one specialist receiving and contributing to an issue before signing it off and passing it to the next, while other matters are managed better if all specialists contribute to a group discussion. Careful consideration must be given to the design and management of communication networks in order for them to function effectively.

Channels of communication and communication networks

There is considerable evidence to suggest that the ability to process information is dependent on the type of network and the nature of the information processing task. Before we look at the strengths and weaknesses of different networks it is useful to identify the main types of communication networks and their character- istics. Familiar communication models developed by psychologists are that of the wheel, chain and comcon (Figure 7.1). The wheel model of communication repre- sents a highly centralised configuration with all information channelled through, or to, one person. While the chain network also includes parties who receive infor- mation from more than one source, no one person has direct access to all the others in the network or receives all of the information. Information flow is up or down the chain and is subject to interference from organisational gatekeepers. The comcon network represents the most decentralised model of communication. All parties in the comcon structure have access to information from all other parties in the communication network.

All these networks are commonplace in construction; the networks can be

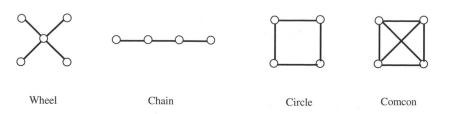

| Wheel | Chain | Circle | Comcon |

Figure 7.1 Basic communication networks.

examined on a small scale, for example, within one organisation, or at a macro level, for example on a construction project where the network extends over organisation boundaries. The wheel provides a useful model to explain much of the formal communication flow during the construction phase. The project manager occupies the central position and the other contributors are to be found at the end of the wheel spokes. The only adaptation needed to this model is to provide two central nodes that represent the architect and the contractor in more traditional arrangements. A model adapted to incorporate this is shown below (Figure 7.2). An advantage of centralised communication networks is that the formal lines of communication are clear, those on the periphery are aware of who to contact for information and decisions.

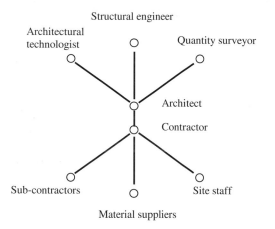

Figure 7.2 Model of centralised network occurring during the construction phase.

Much of the formal communication during the construction phase will flow through either the architect or the contractor. All of the design information is communicated via the architect, and the building and assembly information is channelled through the contractor. Some of the information may travel along long chains before reaching those at the central hub. If problems emerge direct face-to-face interaction may be required to resolve them. The same group of professionals can also come together to interact in a decentralised network such as the comcon. The design team meeting, progress meeting or management team meeting provides a forum where the managers and designers can interact in a way that has greater similarities to what has been called the comcon network (Figure 7.3). During meetings there is potential for open interaction with all members of the groups. There is much greater potential to communicate with other members of the meeting

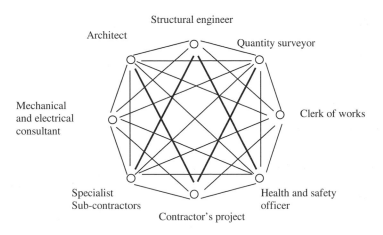

Structural engineer

Architect

Quantity surveyor

Mechanical
and electrical
consultant

Clerk of works

Specialist
Sub-contractors

Health and safety
officer

Contractor's project

Figure 7.3 Schematic of possible interactions in a meeting (meeting network).

than when interaction is controlled through formal networks. One weakness of decentralised networks is that interaction can become uncontrolled, the openness of the interaction can make it difficult to recognise those who control the decision-making process. Figure 7.3 attempts to demonstrate the number of possible interactions in a meeting consisting of eight people. There are benefits to managing communication using centralised and decentralised systems. Some of the issues to be considered are discussed below.

Centralised networks

Decentralised networks

Figure 7.4 Centralised and decentralised networks.

Centralised networks

In centralised networks information flows to the hub of the network. The central person controls communication and can perform the decision-making task alone. The central type of network performs much better than a decentralised network when working on simple tasks. However, when dealing with large amounts information the person at the centre of the network can become saturated and hence overloaded with information. The wheel arrangement does not perform well when working with complex information.

Decentralised networks

Information will flow unrestricted (in theory) to all persons in the communication network. No individual has, or can, control all of the information in the group and so it can be difficult to make decisions. In decentralised networks, the lack of a simple structure can mean that valuable time is wasted on discussions about irrelevant and/or trivial topics or simply on deciding responsibility for an issue when one person could have quickly made a decision. This network does not perform well with simple tasks. When dealing with complex problems in decentralised networks information may come from, and be evaluated by, all of the people in the network. This can be seen in project intranets and extranets. No one person should become overloaded with information. The decentralised network performs better on complex tasks.

Though the structure of the meeting is most closely associated with the comcon, there is the possibility that in the group context within organisation settings some people are less willing to communicate ideas than others. This could change what is perceived to be a free-flowing network into a network that does not necessarily have open lines of communication. Factors such as communication dominance, influence and reluctant communicators may affect the meeting interaction (see Chapter 6).

Communication networks and groups

No group, whether formal or informal, functions properly unless its members can communicate effectively. The free flow of information, knowledge, ideas, emotions and feeling among group members determines, to a large extent, the efficiency of the group and the satisfaction of group participants (Shaw 1981). Shaw (1981) reviewed several studies on the effects of communication patterns on the group process. From the review he has composed a theoretical construction of communication patterns and their effectiveness when solving different types of problems (see Figure 7.5).

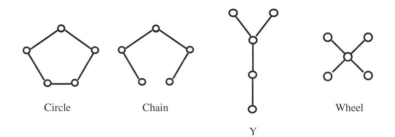

Circle   Chain   Y   Wheel

Figure 7.5   Five-person networks.

Networks and simple problems

In a study of simple problem-solving using a five person circle, chain, Y and wheel respectively, the average times for solving problems were the same, although the single slowest time was the circle, the circle groups made the most errors and Y the smallest. The circle required considerably more messages to solve the problem compared with the other networks. Circle members reported greatest satisfaction and the wheel members the least. Emergence of leaders was also measured. The frequency of a single person being named as leaders was lowest in the circle, then the chain, then Y and was greatest in the wheel. All persons in the wheel experiments nominated the most central person as leader. The operational or organisation patterns observed in the wheel, Y and chain were such that information was sent to the central position and then sent out. The circle showed no consistent patterns of operational organisation.

The consistency with which the leader was nominated in the wheel shows an awareness of roles within the communication network. Knowledge of roles within a group may have positive and negative attributes, for example, knowing that you can only communicate through one person reduces the time spent deciding who to consult. Unfortunately the act of passing information through one person may reduce the effectiveness of individual contributions, since it is impossible to discuss issues with others without going through the central node. The central person has the power to pass on and withhold information, the information going to and from others is restricted and controlled. Knowing that input efforts may be restricted, blocked or changed may reduce an individual's willingness to work to their full potential. In wheel-type models the leader (central node) is the only person who can offer praise or reward for efforts to other communicators, this is one reason why members may experience low levels of satisfaction when operating in such networks. Considerable importance is placed on the central person in wheel-type networks. With less centralised networks, such as the circle, roles and responsibilities are less obvious and would not necessarily restrict participants to leader/subservient roles. In circle-type networks, praise and reward, or blame, can come from more than one person.

Networks and learning behaviour with simple tasks

Group patterns and relationships are important. Communication across the group provides information that is used to develop understanding of the problems, which is part of the learning process. Shaw (1981) reported the findings of information handling, learning behaviour and testing of mathematical problems in communication networks. The number of messages required to complete the task was greatest in the chain and smallest in the wheel. The time taken to solve the problem was fastest in the more centralised networks and slowest in the decentralised networks. In a further study to determine the levels of learning, all networks resulted in increased learning, but only the chain and the circle demonstrated significant learning. The circle groups achieved a high level of efficiency in comparison with the other networks; however, even with learning the performance of the chain was still poor. The major difference in group performance and satisfaction is between centralised (wheel, Y, chain) and decentralised (comcon, circle); the direction of the difference depends on the type of task assigned to the group (Shaw 1981).

Networks and complex tasks

Even slightly more complex tasks can change the effectiveness of networks. Shaw (1981) discussed a small study on experiments involving word and sentence construction problems in three-person networks. The comcon was found to be the most efficient and the wheel least efficient. Shaw (1981) also examined equal and unequal distribution of relatively complex information in communication networks. Rather than giving each person in the network one piece of information, each member of the network would receive different amounts of information. Complex analysis of the result revealed that the circle was the fastest and the wheel was the slowest. The difference between the circle and the other patterns was greatest with unequal distribution of information. No differences were recorded in the number of errors; however, the ability to correct errors (corrective power) was greatest for the circle and least powerful for the wheel network – this was considered contrary to expectations. Leadership was in general agreement with previous findings. Numerous studies have observed that the wheel is faster than the circle for simple

the gatekeeping decisions made by a wire editor on a small newspaper and found that the wire editor (whom he called 'Mr Gates') made 'highly subjective' decisions when rejecting or accepting news stories. Drawing on Lewin and White's research, Westley and MacLean (1957) proposed a gatekeeping model (in which the gatekeeper is also an encoder) that has been used by other researchers. The gatekeeper can select messages transmitted from the sender and pass on to the receiver those messages that he or she feels are appropriate to the receiver's needs. Messages can be withheld or transformed prior to passing them on, thus the receiver will receive different messages to those sent.

Gatekeeping research has largely concentrated on the mass media field with special emphasis on the selection of news items by editorial staff although the construct has been extended by Pamela Shoemaker (1991) and applied to the architect's office (Emmitt 1994, 1997, 1999). Information in the form of messages exerts forces on the individual both from within the firm and from outside the firm. Messages may be deliberately directed at the firm (e.g. building product manufacturer's advertising literature) or exert force simply by being present within the firm's milieu (e.g. social pressures to design for a sustainable future), some may enter the firm, others may be ignored or rejected. From the individual's perspective messages may not be received simply because they have not passed through the organisational gates; furthermore some messages may come directly to the individual, others through the organisation gatekeeper (a modified message) while information generated from within the firm will also exert pressures upon the individual.

Once the individual's behaviour is appreciated it is possible to look at the gatekeeper's control of information flow within a network. Gatekeepers are important since they act as a physical gate through which information has to pass; they are in a powerful position, conveying messages from one person to another or withholding all, or part, of the message. Gatekeepers can help other members of the network to avoid information overload by allowing through only what they feel is important information. Of course, it is the gatekeeper who decides what is important to send on and what is important to withhold, so there is always the danger that information may be withheld that should have been passed on simply because the gatekeeper has made an incorrect decision. It is important to recognise that all members of a network operate as gatekeepers, operating their gates on different levels depending upon their position (perceived position) within the network. The importance of the gatekeeping construct within the project management environment has been made and is addressed later. It is only when the gatekeeping construct is appreciated that it is possible to look at information and knowledge management.

Any person within a chain of communication will act as a gatekeeper; receiving information and deciding whether or not information will be passed on, filtering information, reducing the content of the information or blocking information flow (Shoemaker 1991, Emmitt 1994, 1999). In many cases information will not be blocked on purpose but will be lost or forgotten. Bowen (1995) also introduced the concept of the gatekeeping function into the interpersonal relationship between the client, architect and quantity surveyor. The function of the gatekeeper, in this instance, was to regulate, filter and modify information before passing it on to a third party. Although gatekeepers may block, distort or change information, many organisations introduce intermediaries (gatekeepers) to make important decisions and solve problems. The decisions are often made by more senior professionals when they involve a large redistribution of resources.

Opinion leaders

There will be a number of opinion leaders operating within the network, and these individuals perform a different role to the gatekeepers. As the term suggests, the opinion leaders are concerned less with controlling information as it enters and leaves the organisation, and more with influencing and directing the opinion of people within and without their organisation. They can have a powerful influence on the direction of projects because they will contribute to the project norms.

Levels of construction networks

To understand the opportunity for participation and empowerment in more depth it is necessary to look at the social structure of the temporary project team. Sociologists have argued that any social situation is a sort of reality agreed upon by those taking part in it, thus individuals will have preconceived constructs of what is expected of them and of others in the network. In building projects, the project team exists on three fundamentally different and potentially conflicting levels, namely the formally constituted network, the statutory network, and the informal local participant's network.

(1) *Formal network*   Formally constituted through contracts, the members of the project team are clearly identified and their roles defined. Their aim is to produce a building in accordance with the wishes of the client, usually within stringent time and cost constraints. The formal project network is likely to contain individuals working to different agendas because of their training and position in the network. Interaction with the statutory network and the informal user groups may be sporadic and seen as a distraction rather than a help in terms of the project goals.

(2) *Statutory network*   Various external contributors to the project are represented by statutory authorities, such as the town-planning officer, building control officer, fire officer, etc. These may influence the planning of the project at different times and to varying degrees, through for example, the insistence on the submission of an environmental impact assessment (EIA) by the planning authorities. In the UK this network is usually perceived as protectionist and therefore restrictive in terms of the project. The statutory network will be determined by the physical location of the construction site.

(3) *Informal network*   This will comprise local interest groups concerned with their own wellbeing and/or that of the local ecology. Local groups may wish to participate in the project, through co-operation or through protest (usually by way of the planning authority). User groups may also operate via an informal network. The opportunity for empowerment will depend upon the timing of their intervention and the manner in which their concerns are communicated and accepted by members of the other networks.

Because different networks operate within any one project, it is unlikely that any one individual will have a clear understanding or a complete picture because of the many processes occurring at the same time. Not everyone in the team is linked together, furthermore communication loops within the project are not closed, and thus there is scope for ambiguity, change, interference and misunderstanding. Individuals are likely to have different goals, values and priorities, whatever their position in the network, furthermore each individual will have a unique view of the project based on their own perceptions drawn from the information available to them. No single party is really in control of communication between the networks. However, if effective participation is to be achieved, where everyone involved can get their individual messages and contribution considered, there is a need for a decision-making framework and appropriate leadership. Essentially, the requirement is for an integrated team where all members, whatever network they come from, can provide their specialist input as part of a common goal, a true team that encourages and provides the opportunity for participation and empowerment.

Fragile networks and robust gates

An alternative (complementary?) approach is to model networks according to their function in terms of communication. Wyatt and Emmitt have argued that the dilemma of networking for sustainable design is linked to the diffusion of information and to the networks that form for any temporary building project, namely the social network, the project network, the product network and the information network (Wyatt & Emmitt 1997, Emmitt & Wyatt 2000). The underlying theme is communication, or rather the difficulty of communicating information effectively throughout all stages of the product's life. Information technology is a tool which can assist with the rapid transfer, storage and dissemination of information relating to a project, but this information (and the technologies employed) must be managed effectively if individuals in the project team are to avoid information overload and thus assist, not hinder, the communication process.

One reason why life cycle design was largely neglected in the past lies in the inaccessibility of information – information that exists in the form of specialised knowledge – and which is held by a number of poorly linked (isolated) players in the temporary project network. On the one hand, progress may be achieved through new technologies (IT) making access to information easier and quicker; on the other hand the number of individuals involved in building continues to increase, with the potential to hinder progress. Thus both information technologies, such as computer-based expert systems, and the number of people contributing and drawing on the information within the system need to be managed. Whatever strategies are evolved to secure sustainable building they must embrace all of the contributors to the building process, in both the project and the product phases. Thus building product manufacturers and suppliers, as well as the disassembly sector, must be included within the whole life appraisal if an individual development's environmental impact is to be reduced. We must understand the temporary networks and the challenge posed by an information-driven environment and recognise that the role of professionals is changing. Whether designers, project managers or another construction professional emerges as an information manager remains to be seen. It is important to recognise that information management, not design, is key to the future development and the provision of competitive service provision. This brings us on to the issue of logistics and supply chain management in construction.

Supply chains: strengths and weaknesses

A formalised network of organisations that strategically work together over the longer term with the aim of increasing quality, productivity and profit is known as a supply chain. The development of logistics and supply chains took place in the manufacturing sector and is particularly well suited to process orientated industries, although with increased effort directed at improving the efficiency of construction, many (large) projects have adopted the techniques. In construction the term supply chain management is used to describe the management and improvement of long-term relationships between the client, designers, contractors, sub-contractors and suppliers, i.e. all of the parties that supply the goods and services to realise the building. More importantly, supply chain management attempts to maintain a long-term network of professionals and organisations that work together on more than one project with the aim of incorporating knowledge into new projects in order to achieve continual improvement. Unfortunately, it is the fragmented nature of the construction sector, combined with the temporary nature of projects, that makes it difficult to maintain a degree of consistency between projects. High profile projects that have successfully adopted the concept of logistics and created effective supply chains tend to be very large projects with a high degree of repetition or repeat projects for the same client. For smaller, bespoke project the concepts are more difficult to apply and sustain over the longer period.

Through the use of agreements between clients, designers, contractors and sub-contractors the organisations establish relationships and contracts that enable them to operate together on different projects as a cohesive network. When companies work together for prolonged periods the links between organisations are strengthened as familiarity and trust are built up, thus enabling open and effective communication along the chain. Relationships are developed that seek to improve processes with the aim of improving the delivery of the project. In doing so, operational logistics, communications and process become more efficient and over time services can be improved to increase customer satisfaction, lower costs and improve quality. Through the utilisation of knowledge and a steady flow of work there is more opportunity for organisations to engage in research and development and consider more innovative products and processes. Indeed, it is through such collaborative working that environmentally responsible construction may be realised (see below).

There are a number of downsides also. While the idea of more stable and consistent supply chains is a good one there are some limitations. Organisations forming the links in the chain may remain the same, but employees tend to move jobs and since interorganisational communication relies heavily on personal relationships the reality may be that the supply chain is less strong than it might appear to the casual observer. Furthermore, it is not always clear who benefits from supply chain management because some clients and main contractors have been known to enter into agreements with suppliers and sub-contractors not so much to develop more efficient systems which result in mutual gains to the client and supplier, but with the intention of driving prices down to the bare minimum. Through such links sub-contractors may also become overly reliant on one client: if the client experiences financial difficulties the sub-contractors can also suffer similar problems. There are also serious questions with regard to competition, since once the supply chain has been formed it is very difficult for new organisations to break (into) the chain.

The benefits of strong supply chains are clear. There is greater potential for more effective communication processes when parties can work together making improvements. Parties must recognise that is not the initial agreement, which improves the supply chain, but the way organisations develop and improve the communication processes and information networks that support improvements to services and goods.

Further reading

Rogers, E.M. & Kincaid, D.L. (1981) *Communication Networks: Toward a New Paradigm for Research*, The Free Press, New York.

8 Organisational communication

Businesses that are responsive to internal and external demands have a positive organisation culture in which members are committed to the organisation and who are working together to ensure continued success. The attitudes of employees and the organisation's culture are directly related to the nature of organisational interaction. Competitive advantage is achieved through awareness of market opportunities, the ability to respond to change and the ability to communicate. Clear and timely communication within the organisation and with other organisations and individuals associated with projects is a vital factor in helping to achieve an efficient and profitable business. Here we look at organisational culture, intra- and interorganisational communication and configuration management. This leads into a discussion about the topical issue of outsourcing and e-business in construction. We conclude with some thoughts on innovation and change and the desire to encourage greater collaboration between organisations.

Organisational culture and communication

The importance of effective organisational communication to business success was recognised as early as 1938 by Barnard. Communication was seen to be central to the organisation, with the structure, extensiveness and scope of the organisation being determined by the communication techniques employed. During the 1970s and early 1980s scholars started to question the classical goal-orientated theories of organisation. Critics of the systematic approach to understanding organisation behaviour suggested that, on close observation, organisations are rarely rational. Behaviour is often irrational and spontaneous, may be unpredictable and can sometimes be self-defeating. Rather than following set procedures, management systems or rules, people are guided by, and contribute to, the culture of the organisation in which they work. The organisation's culture and the framework of rules and rituals that it produces are developed through human interaction and social experience. Aspects of culture include member attitudes, values, social rituals, mythology, scientific knowledge, social systems, prejudices, norms, laws, rules, habits, systems, philosophy, hearsay and behaviour. As members enter an organisation they interact and form relationships that help them to explore the cultural ideology of the organisation. Through interaction, norms are observed and accommodated, thus members start to act in a similar way and follow similar decision-making procedures to their colleagues. By communicating with others in the organisation individuals learn the ideas, practices and behaviours that are shared by other employees. It is the shared values and decision-making structures that help individuals to buy into the organisation culture, a characteristic known as 'shared symbolic logic'.

From the perspective of the organisation, the word 'culture' refers to the accepted pattern of a group's behaviour, which includes all of the organisational arrangements, the group's way of thinking, feeling and acting and of course the physical composition of the group (Brown 1963). There are two separate but related schools of thought regarding organisation culture:

- The 'cultural variable approach' regards culture as an influential aspect of organisations. Culture affects the level of conflict, teamwork and performance in the same way as a leader would affect a group.
- The 'culture as sense-making' philosophy sees culture as the essence of organisation providing members with shared interpretations of reality that facilitate their ability to work together and organise tasks.

Cultures that develop within organisations can be either productive or destructive. Many case studies have found that successful organisations have a strong positive productive culture. Such cultures are said to be a primary influence on employees' motivation and commitment. Deal and Kennedy (1982) identified four key attributes of companies that had strong positive cultures:

- *Shared positive values*   The stated beliefs help members to interpret organisational life. Where employees believe in, and adopt, the corporate slogans or mission statements, strong positive values were often found to exist.
- *Heroes*   Where people are influential and personify the strong positive values of an organisation, other employees attempt to emulate the members.
- *Rites and rituals*   These are the ceremonies, acts and events that members use to celebrate and reinforce interpretations about the values of organisational life. Such events enrich and add to the excitement of the organisational activity, making them more memorable.
- *Cultural communication networks*   These are the informal channels of interaction that are used for indoctrinating members into the organisation culture. By exchanging information, sharing jokes, recounting stories and discussing legends the positive values of the organisation's culture may be reinforced.

Informal conversation within organisations has been found to be an important process for understanding what were considered as 'taken-for-granted' statements. Conversation was found to be essential in overcoming ambiguity, enabling the organisation to function effectively and hence be productive (Hosking & Haslam 1997). Pietroforte's (1992) research on management systems used in construction projects found that while projects were supposedly governed by formal contracts, the decisions made in projects were based on informal relationships and roles. The procedures used were more about the 'accepted' methods of working than about the prescribed contractual procedures. It follows that individual projects will develop their own culture, formed through the interaction of different organisational cultures, and so this, too, needs to be managed in order to develop a culture that supports the transfer of design intent into a physical product, a point taken up in Chapter 9.

Culture and the role of education and training

To develop shared beliefs, create heroes and reinforce positive values through rituals, many construction organisations are now rewarding their most outstanding employees by publicising and publicly praising their performance. Such reward mechanisms are not only concerned with the achievement of targets and goals, but also for showing some creativity and working in a more innovative manner. The desire to improve working methods and change organisational culture to reflect changing market conditions can be achieved through educational and training schemes. The use of in-house and external presentations, workshops and inde-pendent-study, if developed and planned correctly to suit the needs of the orga-nisation and the individuals concerned, can reinforce positive cultural values and also reinforce the value of specific learning practices.

Organisations may also attempt to change or reinforce attitudes by sending let-ters, memos and other literature to their staff via electronic and paper-based media. Yet it is well known that many employees do not read all of the information sent to them, simply because they do not have time to do so. Changing or establishing cultures is difficult and requires considerable effort from employers and employees alike. Best practice and new ideas must be discussed and debated before imple-mentation, and this relies heavily on interpersonal interaction. Social events can help stimulate informal exchanges, improve internal relationships and instil a sense of company community. Once again, the point needs to be made about aligning events to the culture of the organisation and providing the opportunity for all members to take an active part, thus helping to develop and reinforce informal interaction.

Education, training and development programmes may help to improve the confidence and skill level of the organisation's members; however, the point needs to be made that employees will only believe in the company values if and when good practices are adopted and performance is improved. We need to know when the strength and the performance of the company has improved and/or is better than that of a competitor and we can only gauge this by some form of measurement. Construction organisations are now adopting benchmarking techniques to measure their performance against specific targets. While this can have a positive effect on morale there is also the danger of employees becoming despondent if targets are not met. Obviously, the results of a benchmarking exercise need to be reported accu-rately (or why bother?), but more importantly the whole issue needs to be carefully managed and feedback on performance discussed with employees by way of face-to-face interaction. Managers should also be careful not to overemphasise minor achievements when it is clear to employees that the 'bigger picture' is less healthy, which can make employees cynical and distrusting of their managers.

Implicit exchanges of information

The organisation's culture is developed through interaction. Messages are exchanged about the organisation and its practices. Some of the information will be clear and explicit, other messages, while still powerful, may be implicit. When new employees enter an organisation they will receive feedback from their peers about whether their actions are acceptable. Facial expression, body language, emotional and emotive communication, such as others turning their backs, frowning, praising, laughing, expressing support and showing tension, will send signals to the new-comer. As individuals become accepted into the organisation other members will engage in informal conversation. A method of communicating cultural information that is commonly referred to is story telling. Exchanges of stories about the orga-nisation, whether based on fact or fiction, can create positive and negative images of the organisation. Jokes about employees and the organisation also convey powerful and persuasive information. Digs and jibes about a manager's mistakes help to let others know about the manager's ability and performance and individuals can take action to mitigate the manager's idiosyncrasies.

Inducing members to positive company values

To ensure that new employees are introduced to the positive company values it is essential that their induction period is supported by positive cultural interaction with colleagues. Many organisations have now adopted the 'buddying' system where new employees are teamed up with individuals who will help them with problems and explain the company systems. The role of the buddy is clearly important. Only those who are supportive and portray a positive company image should be used in this role.

Balance of individual and organisational goals

The organisation's culture will influence the way individuals and organisations interact to achieve their goals. Organisations can be differentiated in terms of the way its members and the organisation's objectives are balanced and managed (Barrett 1970). The three types are:

(1) *The exchange model*   This approach is based on bargaining related to organisation goals and individual rewards. No attempt is made to integrate organisation and individual goals, instead a relationship between the individual's activities and the organisational goals is developed that helps and then rewards the individual on achieving set tasks. It can be viewed as the balancing of inducements and contributions, or the exchange of a working currency that focuses on social incentives.

(2) *The socialisation model*   This system pursues the integration of organisational and individual goals through persuasion or social influence. Employees and members are encouraged to value organisation tasks and activities and contribute towards the achievement of organisation goals.

(3) *The accommodation model*   This model focuses more on individual goals. The approach takes a greater interest in the individual and looks at how the organisation can satisfy the individual yet still achieve organisational objectives. It is the organisational procedures that change to accommodate individual needs.

Recognising that individuals work for many different reasons, apart from assisting the organisation in the achievement of its goals, is important. Failure to provide members with the appropriate level of remuneration, reward, professional development and social recognition may result in poor performance and/or low levels of satisfaction. It is clear that individuals have different needs and respond differently to the various management strategies. Some will prefer to know exactly what they are required to do and what they will receive for their performance, others respond better when placed in empowered positions where the methods of work, targets and rewards are more flexible. In order to balance and maintain individual and organisation objectives, managers must constantly interact with employees checking levels of satisfaction and performance.

Motivation

Productivity and quality are related to the degree in which the individual is 'engaged' and committed, and is closely linked to delegation (Maister 1993). Personal motivation is a complex area, but one well covered in management literature. Motivation theory is based on the fulfilment of an individual's needs (e.g. Maslow 1954, Hertzberg 1966) and the manager who is able to motivate and reward staff fairly is well on the way to establishing an organisation with competitive advantage. For example, a good designer may need little motivation when designing, but may need encouragement when dealing with a part of the job, such as contract administration, which is seen as non-creative and therefore less interesting. Managers must understand what motivates the individuals in their organisation, and this is not always clear. Formal staff review procedures are in place in the majority of organisations with the aim of discussing and identifying the needs and aspirations of the individual and matching them to the needs and aspirations of the business. Such mechanisms are useful; however, it is only by watching and listening to employees that their level of motivation becomes obvious. Thus there is a need for managers to engage in more informal conversation with employees to better understand them (and for the employees to better understand the motives and actions of their manager).

Managing conflict

The management of conflict within organisations needs to concern itself with the reduction and eradication of dysfunctional conflict and the encouragement of functional conflict with the aim of enhancing creativity. The hierarchy of an organisation is often seen as a way of resolving conflict. Where two or more individuals, or groups, within an organisation are unable to resolve differences a senior manager may intervene, with individuals and groups being more likely to accept a decision made by someone at a more senior level. Where two or more organisations are involved in a problem which is not resolved by those directly involved, more senior representatives from each organisation may need to intervene. A certain amount of diplomacy is required on behalf of those intervening to avoid resentment and ill feeling.

Many mangers believe that conflict reduction is achieved through adequately informing and involving their employees. Some are often disappointed, however, to find that with increased involvement and information comes the potential for greater conflict. As more people are involved in decision-making the potential for disagreement increases. Although the mass of disagreement may be greater with increased numbers of people, the intensity of the disagreement or conflict may not be as great, or as polarised, as that which is associated with smaller decision-making groups such as the dyad. Conflict resolution becomes more problematic when associated with different organisations contributing to the same project. Organisations may seek to secure their own goals before addressing those of the temporary construction project. Partnering procurement methods have attempted to introduce win–win methods into the construction process. This attempts to remove a company's desire to make gains at the expense of other organisations.

Intraorganisational communication

Intra or internal organisational communication occurs within the boundaries of a company, i.e. among its members. When communicating within an organisation, where no external members are present, it is clear that communication is limited to internal messages. In some project environments people from different organisations will work together. Some discussions with colleagues from the same organisation will be overheard by, or may also be intended for, project members from other organisations, which is classed as interorganisational communication. We can only consider communication to be truly internal when it is limited to those operating within one organisation.

Internal communication uses both formal and informal communication channels. The formal channels are clearly planned and established by the organisation. Formal lines of communication within an organisation are either hierarchical or lateral (between departments). The formal lines of communication help identify and create hierarchical levels in an organisation, divisions, departments, teams, positions, responsibilities and roles. Formal communications provide the organisation members with a framework of information, yet in most cases the formal communication channels do not satisfy the needs of the individual employees. Informal communication channels emerge and develop through social interaction and can be used to help members receive information needed to perform their duties. Both informal and formal communication can have positive and negative effects if not controlled. The potential for managing interaction within one organisation is far greater than controlling interaction in projects where more than one organisation is involved. As already discussed, it is essential that the formal and informal communication systems are supported by a positive organisational culture. Positive cultures help increase morale and performance internally and send positive images to those outside the company, for example, if those employed create a positive company image, this is communicated to potential employees, clients, contractors and competitors.

Communicating quality

For the purposes of this book we are largely concerned with the quality of the communication exchange between organisations and individuals party to a particular building project. Quality management systems and communication are inherently linked because without clear communication the quality management systems will not operate effectively and may have a detrimental effect on employees and project outcomes. What we need to look at is how quality management systems may help to improve communication with a view to reducing the potential for misunderstanding and possible conflict.

There is somewhat of a paradox about quality management systems, since the procedures adopted in order to improve monitoring and control often result in more paperwork, not less: a constant complaint of those who work within a poorly designed system. Many organisations within construction have questioned whether quality assurance/quality management (QA/QM) is worth the investment, and critics have claimed that QA is little more than a form-filling exercise, that it is a fashion that will pass, or is just a marketing badge to attract clients. But they have overlooked both the demand from clients and the benefits that QM systems can bring to the project and hence the finished product. At the heart of QM is the total commitment of all the organisation's members to quality, total quality management (TQM). Commitment comes from initially raising staff awareness about QM through internal communication within the organisation, through specialist training for both the quality manager and the auditors, to general training and updating. Central to the quality ethos is communication and especially feedback. Each individual must make the aim of continuous improvement central to all their activities and feel free to contribute to the process through feedback to the senior management via feedback mechanisms such 'quality circles'.

Quality circles

As a result of investigations into Japanese management techniques, many American businesses have adopted the concept of quality circles. Recently the use of quality circles, also called circle-time, has become common practice in British organisations. Quality circles are meetings that normally consist of a small number of people, ideally between five and ten. The meetings are used to collect ideas, identify, analyse and openly discuss work-related problems. Thus, those who attend the meeting normally have a common interest; members of the group either work together or undertake activities that affect others in the group. Those who volunteer to take part in the quality circle normally receive training. The training helps members engage in sensitive discussions and identify problems without the issues developing into personal confrontations or major disputes. While quality circles are rarely used on construction projects, occasional reports on their use and usefulness are emerging. Hall (2001) used quality circles to identify the reasons why particular problems occurred on a construction project. The method proved to be particularly good at identifying common problems experienced between members of the construction team. Identifying problems both during and at the end of project increases the potential of preventing such difficulties re-emerging in the future.

Controlling what we give, when and to whom

Furthering the debate about information overload and information transmission is the wider issue of company policy. Whatever the legal constitution of the organisation there is a legal obligation for the business to comply with prevailing legislation, both UK, European and worldwide. Such legislation is interpreted and written into company policy documents, many of which are very extensive, time-consuming and also demanding to read. This raises two fundamental questions. First, do the employees have time to read and understand the policy as it relates to their area of work? Second, do people really understand the implications of what they are reading (especially since much of the writing is in 'legalese', a language lost on most of us without a legal background)? Readers will find a lot of surplus information, and what makes sense to one may be completely irrelevant and incomprehensible to someone else with a different job function. So we will select information as and when we need it. Do organisations really provide enough time for their employees to read and understand their policies? Of course the senior managers would argue that they do, but most employees with increasingly diffi-cult-to-manage workloads would argue otherwise. A further issue relates to what organisations are telling their customers. Are they doing what they said they would? And if they are doing something different, is it necessarily a problem? Have they remembered to keep the client informed and discussed the consequences of the changes? Customer expectations need careful handling. In both situations descri-bed above we have to ask the question: what happens when things go wrong? The, rather obvious, answer is that people turn to the written policy documents for comfort and reassurance, often to find that they should have consulted them earlier.

Interorganisational communication

Construction project communication is typified by interorganisational relationships. The designers, engineers, managers and skilled workforce belong to different organisations. Each employee and organisation has its own objectives, yet for projects to be successful the effort of those involved in the process must contribute towards the project goal. The co-ordination of activities relies on the effective formal and informal communication practices. Each organisation operating within the project will have its own formal and informal communication procedures. If communication over organisation boundaries is to be effective, the formal and informal communication practices of those involved in a project must have a strong influence on the individual organisations. The lead organisation in the project team must develop a communication system that other organisations can adopt and with which they can work. Both formal and informal project systems must provide clear guidance, yet be flexible so that individual organisational communication practices are not compromised to the extent that their performance is hindered. The contracts, management systems, meetings, letters, emails and conversations provide the control tools of the organisation, each must be tailored to meet the project needs.

Construction contracts establish the formal relationships and responsibilities of those involved in the construction projects but rarely govern the communication procedures and management systems. Contracts provide the formal basis of the procurement process, setting out the terms and conditions, roles and responsibilities and allocating risk. Once contracts are exchanged each party will seek to minimise its contractual risk by managing and controlling the project to the best of its abilities. To ensure that others, whether directly or indirectly employed, perform their duties, all communication channels must be managed. The combination of communication systems used to manage projects should be designed to:

- Induce new participants
- Inform and agree action
- Identify targets
- Allocate responsibility
- Measure and control performance
- Co-ordinate information and activities
- Check and remind others that critical activities are in place
- Encourage, persuade and enforce action and behaviour
- Identify and resolve problems
- Re-schedule activities
- Provide feedback on client satisfaction
- Manage and resolve conflict
- Negotiate and control disputes.

Planning a system that will set up formal communication practices and induce informal interaction that supports project objectives is essential. A considerable amount of work has been undertaken in the field of internal management communications (for example, Hargie et al. 1989; Kreps 1989, Rasberry & Lindsay 1989), however, little work exists on communication in construction that is exchanged between individuals in the same organisation and over organisational boundaries. Some early research work by Gameson (1992), Pietroforte (1992), Bowen (1993), Loosemore (1996), Emmitt (1997), Hugill (2001) and Gorse (2002) has started to develop an understanding of the nature of such interactions. The chapters on groups, networks and meetings deal with some of the issues raised in their research.

Systems theory and internal and external interaction

The systems approach to management has considered the differences between the closed interaction system of a single organisation and that of an open system where organisations are not considered in isolation. All organisations exist in an interactive environment made up of other businesses and customers that fluctuate as a result of economics and politics. Organisations and their environments are interdependent – as environments change, organisations must respond and shape their milieu. The development of a new product can change, the nature of construction. The development of the prefabricated trussed roof rafter has revolutionised the way roofs are constructed. The vast majority of new houses are now constructed with trussed rafters rather than more traditional forms of roof construction. While a company's development of the trussed rafter changed house construction, much of the prefabrication we see in buildings today is a result of the shortage of skilled trades available and clients wanting shorter construction times, thus the environment is affecting and changing the way we manage and build. When considering organisations using the open systems theory model it is clear that external communications are just as important as internal communications.

Configuration management

A formal system for managing information that has been adopted by the project management fraternity is that of configuration management. Configuration management was originally devised by the engineering industry to monitor and control the assembly of manufactured components. Because it is common to develop various versions and adaptations of the same product it is essential to know which components and pieces of information relate to the adapted products. The process ensures that every component and piece of information is centrally controlled. Before information is released it is checked to make sure that it is correct and will function and fit with other products and processes in the system. Configuration management aims to ensure that all information is fully integrated and easy to access and use. The use of configuration management techniques is particularly suited to the construction industry due to the number of organisations and parties involved who create and share vast quantities of information, and this needs to be centrally controlled. Adopting simple configuration management procedures can help to reduce conflicting information and the associated wasted work. The main components of configuration management include:

- *Central control*   The distribution and storage of current and archived information is centrally controlled
- *Release control*   No information is released without central authorisation
- *Status control*   It is clear whether information is for development purposes, has been approved for use in production, or has been superseded
- *Responsibility*   It is clear who is responsible for tasks and information
- *Distribution*   It is clear where and when information is to be distributed
- *Change control*   All changes are controlled centrally, each change suggested will be checked by appropriate specialists, an assessment of the impact of the proposed changes on other products or systems is made and, if necessary, changes are released, implemented and controlled
- *Integration and verification*   On multi-disciplinary projects, interface meetings between specialists from the various disciplines are used to integrate components. Before changes can be made each specialist needs to assess the impact of the change to ensure conformity, ensuring that the final product fits together and functions
- *Quality management*   As all documents and processes are stored and controlled centrally, the documents can be retrieved and processes checked to ensure they are, or were, undertaken correctly, i.e. the process can be audited.

Some large projects will, by their very nature, have very complex configuration management systems; however, the key principles will be retained. They are simple: collate centrally, assess, integrate, control, store and release. Figure 8.1 illustrates an outline of a simple system.

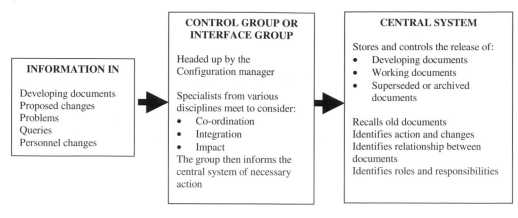

Figure 8.1   The main stages of a simple configuration management process.

International standards of configuration management

There are a number of documents that provide guidelines on configuration management. The codes of practice are concerned with specific application e.g. *BS code of practice for Configuration Management of computer based systems, BS 6488:1984.* There is also a generic framework, *ISO 10007 Quality Management – Guidelines for configuration management,* which may be more informative and useful in helping to implement and align configuration practices that others may use on both a national and international level.

The use of Internet or Web-based document control systems vastly increases the potential for implementing configuration management systems. Drawings and information can easily be controlled from a central node and information can be accessed relatively easily and quickly. If some information is sensitive, the levels of access can be controlled through the portals: a few members may be allowed to access archived material, information on proposed changes, impact assessment meetings and configuration discussion groups, etc.

Outsourcing: a communication challenge

The main contractor will undertake overall responsibility for the construction of the building, but most of the work will be sub-contracted (outsourced) to smaller, more specialised contractors (see Figure 8.2).

The network of contractors, sub-contractors and other bodies involved in major projects is often very complex. Historically, the management of the chain of suppliers of services and products has been difficult and received little attention. With each new project different organisations are involved, all of the stakeholders go through a new learning curve, attempting to understand the management systems, informal processes, design information and products used by the various companies. Although those at the head of the supply chain should have control of the products and services further down the chain, in practice this can be difficult to achieve. Greater emphasis is now placed on the management of the supply chain. The building supply chains consists of a series of activities and processes involved in the transformation of raw materials into a final product that is purchased by a client. On each project the network of supply chains tends to be unique. Clients and major contractors are seeking to develop a more consistent set of suppliers whose

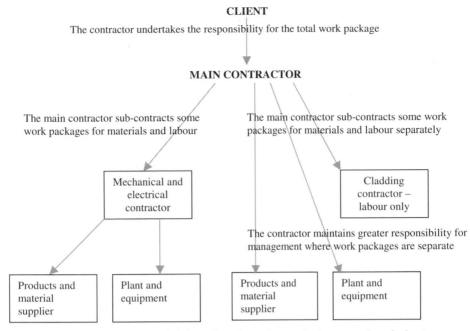

CLIENT

The contractor undertakes the responsibility for the total work package

MAIN CONTRACTOR

The main contractor sub-contracts some work packages for materials and labour

The main contractor sub-contracts some work packages for materials and labour separately

Mechanical and electrical contractor

Cladding contractor – labour only

The contractor maintains greater responsibility for management where work packages are separate

| Products and material supplier | Plant and equipment | Products and material supplier | Plant and equipment |

The subcontractors may also break their work packages down and sub-contract them further

Figure 8.2   One level of a supply chain.

services can be repeatedly used on different projects. The repeated strategic use of contractors and sub-contractors can help to improve communication links and increases the degree of control.

The development of long-term relationships increases the potential to:

- Understand each other's needs
- Develop stronger relationships
- Improve accuracy of pricing
- Improve processes and logistics
- Resolve recurring problems
- Improve quality
- Increase efficiency
- Become more competitive
- Secure work

Some major clients and contractors are entering into agreements (such as partnering) with contractors, subcontractors and suppliers where the parties state that, wherever possible, they will work together. Often to maintain a degree of competition clients will work with more than one contractor. In order to build up relationships and maintain a degree of consistency only a few specifically identified contractors are invited to tender for each contract. Such strategic alliances help to improve and stabilise the supply chain.

Global markets and supply chains

Although there are moves to create more consistent supply chains, the pressures of the global market may threaten established business links. The increased use of e-business and the ease with which services and products can be supplied from various parts of the world vastly increase the number of potential suppliers. Future markets will become more competitive. While more established and consistent supply chains will help efficiency and improve the potential information exchange, global economics may mean that other suppliers, in other parts of the world, become more competitive and can provide an equivalent or better service. Lower wages, strength of the local currency, local skills and expertise all affect the sale-ability of a product or service.

E-business in construction

Face-to-face interaction and human decision-making is essential if we are to respond to implicit, intangible and unpredictable elements of business. However, there are many formal interactions that can be improved by means of digital data exchange and simple computerised decision-making. A considerable amount of the construction supply chain integration can be processed more cost effectively using Internet and Web-based platforms that link multiple organisations. Many other industries have already adopted automated electronic processing of information and data exchanges. For example, as the checkout assistant scans our goods through supermarket tills, the level of stock is automatically adjusted, orders automatically sent to suppliers and our bill calculated instantly. All of this is achieved with little human involvement. As supply chains are used to procure goods and services for an organisation's core business activities (Millet *et al.* 2001), large gains can be realised by adopting electronic business practices to manage data exchange between organisations in the construction supply chain. The intranet and Web infrastructure are important worldwide channels for e-commerce and business-to-business transactions. Although information about construction services, products and tenders are accessible through the Internet, the potential of e-business remains unrealised in construction. Ribeiro and Lopes (2001) identified some types of e-business that are currently being used in construction supply chains:

- *Business-to-business (B2B)* Electronic and online tendering, bidding, surplus auctions, procurement planning, requests for quotations, cataloguing, information exchange, project management, virtual enterprises and banking
- *Business-to-consumer (B2C)* Electronic and online retailing, consulting, real estate, servicing training
- *Intra-organisational* Enterprise workflow, co-operative design over networks, managing and sharing documents and drawings, online meetings
- *Consumer-to-consumer (C2C)* Online actions and services.

Specific processes that would benefit from e-business included:

- *E-procurement* Procuring projects, components, plant, services, experts and manpower; disseminating and collating information about projects, components and services
- *E-commerce* Transactions between the buyer and seller in the supply chain
- *E-logistics* Delivering parts, components, materials, plant and information to the point where they are needed
- *E-collaboration* Co-ordination of decisions and activities among supply chain partners; collaborative design, planning and project management; information exchange between business partners, such as orders, invoices, plans and specification; configuration management and change management systems

- *Customer self-service*   Online technical assistance, training, or guides available for users and customers
- *Auctions*   Auctions of assets, parts and components.

The use of e-business may help to remove, simplify and/or speed up some of the management systems used in construction; indeed, the effect that e-business will have on construction over the longer term will be interesting to monitor.

Performance management

It is essential that employees and organisations know that what they are doing is correct and meets targets and that they are aware of how they are performing against others. Waiting until the end of an activity or project to find that the level of performance was poor or unacceptable, is often too late. Most activities can be quantified and measured at intermediate stages, and key performance indicators provide a method of benchmarking practices against national trends (see also Chapter 9). To measure performance, activities must be broken down into their component parts. It is important that measurement processes are realistic and not dissected into so much detail that measuring becomes too onerous. The performance measures should be easily obtainable with minimal additional effort. Factors that can be easily measured include:

- Timing of activities – start time and completion time
- Budget – measured against cost of work packages
- Number of resources deployed – against those planned
- Safety practices – number of hazards and accidents reported
- Productivity – achieving targets
- Quality – defects reported

Although not dealt with here, the 'earn value management' technique is particularly useful for reporting and measuring costs, time and value together. Most comprehensive project management texts cover this topic.

Using information to measure contractor and sub-contractor performance

It may be necessary to gather information in order to monitor and report progress. Before doing so, however, it is useful to look at the information generated through construction projects. There will be a considerable amount of information that has been captured through quality management systems, health and safety procedures and sub-contractor documentation that can be used to measure performance. In the majority of cases this information will provide the data required for measuring performance.

All sub-contractors go through a procurement process where they agree to undertake specified work. During the procurement process sub-contractors will provide a package of information in which they will identify the quality of their workmanship, the standards to which they work (International or British Standards), codes of practice that they follow, safe methods of work (method statements), short-term programmes and schedules. If the package of work has been evaluated and examined correctly during the procurement stage, this information should then be used to measure and report on the sub-contractors' performance. It is clear that there is insufficient time to watch over every person involved in the

thing to be welcomed or a bad thing to be resisted depends on the individual, the culture of the firm and the environmental conditions prevailing at the time – it depends on one's perspective and one's perceptions.

Kotter and Schlesinger (1979) have provided a 'contingency approach' to overcoming resistance to change, claiming that the managerial approach should be 'contingent' with the degree of resistance. Their six strategies – communication, participation, facilitation, negotiation, manipulation and coercion – range from the least powerful, communication, to the most powerful intervention, coercion. Not surprisingly, the authors argue that communication and participation are the two preferred approaches, recommending manipulation and coercion in only the most extreme of cases.

Understanding causes

It is one thing to propose a series of managerial techniques to overcome resistance to change, but we need to take stock for a moment and consider the causes of resistance. This can only be achieved by looking at individual organisations and/or projects to identify the areas of resistance; i.e. it calls for research into organisational and project culture. Literature on resistance to change is largely concerned with organisations. As discussed earlier in the book, the challenge in any building project is that there are a whole host of organisations and individuals associated with any one particular temporary building project. If, for sake of argument, we consider the temporary building project as a supply chain, resistance within any one link of that chain may affect the project. Classic examples are the piecemeal adoption of quality assurance and the slow uptake of an environmentally responsible approach to both design and construction. These are innovations that have been slow to diffuse within the construction sector.

Encouraging collaboration between organisations

Production industries have introduced systems such as TQM and concurrent engineering in an attempt to improve communications, quality, productivity and customer relationships through closer co-operation. These methods are being tried in the construction sector, with concepts such as partnering and alliancing becoming increasingly popular. While this clearly has benefits over an adversarial system there is no room for complacency. The number of links in the supply chain has not been reduced, simply stabilised, and so the potential for poor communication between intermediaries still exists. It would appear that improvements are linked to the degree of collaboration, integration and synergy that is developed between organisations, and this in turn affects the ability of those contributing to the temporary project culture to maximise their collective knowledge. So the argument has to be for greater simplicity in the design of the project team, with emphasis on a shared culture and collective responsibility. It follows that clear leadership is paramount to the effectiveness of the project.

An integrated approach to design and construction offers the potential for implementing the ethos of concurrent engineering. Whether this is designer- or contractor-led will still depend on the type of project proposed and the aspirations of the client. What is important is that the number of links in the product information chain is reduced and the opportunity for continuous feedback and continuous evaluation of design and production is facilitated. Improved feedback, so often lacking in the construction process, will have implications for durability, will provide the opportunity to look at the total product life cycle and may go some way to engendering a sense of ownership in all of the project's participants. The biggest advantage of a simplified procurement route is the ability of user groups to identify the project co-ordinator and get their views heard, simply because there are potentially fewer barriers to their messages. The project co-ordinator must have the responsibility and willingness to invite participation from the local community via their specialist interest groups at the start of the project, i.e. *before* design commences. It is then, and only then, that the specialist knowledge of the participants can be incorporated into the design process. With careful planning, such participation may not necessarily add time on to the project programme, indeed the potential is also there to save time through the use of expert local knowledge. Early participation has another advantage in that differences of opinion can be discussed early in the project before potentially destructive conflicts of interest develop.

Further reading

Hargie, O.D.W., Dickson, D. & Tourish, D. (1999) *Communication in Management*, Gower, Hampshire.
Kreps, G.L. (1989) *Organisational Communication: Theory and Practice*, 2nd edn. Longman, New York.

9 Building an effective communication culture

Construction project managers occupy a powerful and challenging position from which they can design the project culture, control communication, and hence determine the effectiveness of individual projects. It is the project manager who is responsible for organising systems and resources that facilitate the translation of design intent into the physical reality and who will affect performance parameters such as cost, quality, safety and time. In this chapter we look at communications from the perspective of the project manager, i.e. managing communications, information, knowledge and people with a view to building an effective communication culture.

The construction project manager

Project management is concerned with the achievement of the project objectives using human and material resources, within the context of the project environment and also within a defined time period. This involves the planning and monitoring of tasks, although it is the people involved in the project that will undertake activities and hence operationalise the process. The realisation of activities requires interaction between the parties responsible for the various work packages. Without interaction the plans, programmes and schedules will not develop into real activities that achieve the project objectives.

Project management

Before looking at some of the issues concerning project management, it is necessary to state the obvious. First, all projects are unique, in that each differs from that preceding it. Second, the project is a temporary task for the project participants. Thus not only do the site, product, objectives and application vary between projects, so, more importantly, do the project participants. The uniqueness of a project means that project managers will be faced with a different set of circumstances from previous projects, which in turn requires the formation and maintenance of a new project 'team'. To achieve the end goal within the desired timeframe, the project manager is tasked with the assembly and co-ordination of resources and activities. Early work into project management tended to focus on project management techniques and tools to improve project delivery, indeed it is not uncommon to find this is still a primary concern for project managers in construction. While the effective application of project tools is still necessary, the focus of project management has moved to the people involved in the projects. Regardless of the manner in which the project has been planned, whether it is judged a failure or a success will depend upon the individuals involved, hence getting the correct combination of people is crucial. The assembly of the project team is just as important as the briefing process, since the culture of the project will be set by the people involved and their interaction during the project.

As different parties (consultants, contractors and sub-contractors) enter the construction process their involvement must be managed. Prior to commencing any works an initial meeting between the project manager and party is useful to identify work packages, develop relationships, introduce people to project systems, documentation and procedures, all of which will help to contextualise the working relationship. Early one-to-one briefing will assist the project manager in the development and maintenance of the project culture. Meetings, discussion and instructions will also be required throughout the project to maintain the project culture and help to achieve the project goal. However, the temporary nature of project teams and the way that personnel enter and leave the process at different stages make it difficult to encourage parties to buy into the project culture. However, it is evident that each project has its own culture and own informal and formal leaders that help create the project culture. The task of the project manager is not to allow an adversarial culture to develop that will hinder the performance of the project team, but to foster a culture that instils a positive attitude towards working with others and completing tasks. Debate as to who should be the project leader (architect, entrepreneur, project manager, etc.) is less important than the skills of the individual and his or her organisation to carry the project out. The challenge for the project manager is to recognise and design the communication networks that are likely to be the most effective for a given project, manage communications during the project (to minimise the effect of communication breakdown) and provide the necessary guidance to the project participants throughout the project. Thus, project managers occupy a challenging and powerful position. They determine the procurement route and put together the project participants, in doing so they help to shape the project culture and the project communications, i.e. they will have a major influence on the success of the project.

Who makes a good project manager?

The dilemma as to who should manage the project remains an emotive topic and one difficult to discuss without the issue of leadership and fragmentation rising rapidly to the surface (Emmitt 1999). But we need to address it in terms of communication within the temporary project structure. Certain specific skills are more important than the individual's background and qualifications. We should not get too hung up over who best suits the role, be it an architect, technologist, engineer or construction project manager. The following example is taken from the experience of one of the authors of two very different project managers. The projects were of a similar size, of a similar value and both were run as management contracts.

On the first project the project manager used a rather autocratic managerial style. Communications were tightly controlled and the only opportunity for individuals to discuss problems occurred at the monthly site meeting. The site meetings were chaired and conducted in a professional manner. However, when the minutes were issued they appeared to have little relevance to the issues discussed. The minutes painted a rosy picture of the project's progress; problems were not reported in the minutes. From the second meeting onwards both the architect and structural engineer took their own minutes and circulated them in addition to the project manager's minutes (much to the annoyance of the project manager). In the progress meetings problems were not debated adequately and the project manager's habit of blaming a consultant for the problem before discussing what had happened did little to enhance relationships between the project participants. This project was handed over to the client on time, but with a number of unresolved issues for which no one would take any responsibility. The ensuing dispute took just over twelve months to resolve with the project management organisation putting right the errors at their cost, yet still refusing to accept responsibility for them. To the best of our knowledge the client did not use any of the consultants again.

On the second project the project manager was less autocratic. Communications were channelled through the project manager, but in this case the vast majority of information was disseminated to the whole team. Individuals were encouraged to discuss issues and the whole project progressed with a sense of ownership and commitment. This project was handed over on time and slightly over budget to a satisfied client. This resulted in repeat business from the client and further, equally successful projects for the various consultants. After five projects the individual consultants were operating more like a team than a series of groups, formal communications were used less than on the first project and problems were dealt with quickly and informally because a degree of trust had developed (held together by the project manager's skills). Issues were dealt with using informal communication networks (the problem for researchers looking at this project in hindsight is that there is very little evidence of this, none of the problems are recorded in the correspondence/minutes, etc.).

What may surprise some readers is that the project managers were of a very similar age (late 30s), male, well qualified, experienced and both working for the same organisation – a highly respected firm of project managers and cost consultants. Yet their interpersonal skills and approach to the management of their respective projects were completely different. One was far more effective at delivering a successful project than the other. Analysis of the projects revealed the following facts.

The first project manager operated a closed approach to communications, only passing on that information he deemed to be of value to other participants (much to their frustration). When problems arose he responded by sending a written communication to the consultants he thought to be responsible, requesting their immediate response. There was no attempt to engage in interpersonal communication. This project manager lacked the vital skill of being able to communicate on an interpersonal level; more specifically he lacked tact and diplomacy. The second project manager operated a much more open approach to communications. In comparison with the other manager, he made more information available to the consultants. When problems arose (and several did) this project manager called a short meeting on site for everyone to attend, air their differences and then take a decision so that the project could proceed, i.e. he used the forum of the meeting to good effect. Although these informal meetings took up additional time and were called at short notice (which we would argue is not particularly good practice) the strategy eliminated a lot of correspondence, prevented any differences getting out of hand and allowed the project to proceed to a successful conclusion. He appeared to have a better appreciation of how people behave and communicate and was able to manage the situation far better than his colleague.

Desirable skills and attributes

Given their role in the project, the project manager must possess a wide variety of skills and knowledge in order to perform effectively and thus help others to accomplish their tasks safely, to the correct quality, within budget and within the agreed time-scale. Building partnerships and alliances to enable an open communication culture to develop, based on mutual trust, and which allows groups, and hence the project, to achieve its goal, is fundamental. The ability to understand and manage people, co-ordinate and manage information, together with a thorough knowledge of building design and construction technologies are key skills. Obviously, leadership skills are paramount (see Walker 2002) and in regard to effecting a dynamic communication culture the following skills and attributes are particularly important. These are, the ability to:

- Provide strong, clear and consistent leadership
- Develop empathy with all contributors
- Communicate effectively within and between different levels
- Develop good relationships with informal leaders
- Recognise and manage organisational gatekeepers
- Compose and hold disparate groups together
- Encourage intergroup communication
- Incorporate feedback, thus keeping all informed
- Arrange and chair meetings
- Deal with crises quickly and openly
- Manage conflict to the benefit of the project
- Benchmark project performance
- Establish the project attitude to risk
- Establish the project attitude to innovation
- Communicate and reinforce project goals
- Communicate and reinforce safety information
- Communicate and reinforce quality standards.

Key performance indicators

It is impossible to determine how successful a project is without some form of scale on which to measure. Construction organisations have started to adopt benchmarking tools such as key performance indicators (KPIs) to help them to measure their performance in designated areas against industry standards. It follows that the project manager must consider these indicators of performance while designing the project culture, essentially using the indicators and the knowledge gathered from benchmarking previous projects to build an effective project culture. Benchmarking may be carried out to compare performance against internal operations, against specific competitors' performance or against generic benchmarks that apply to a wide range of businesses.

Construction industry KPIs are national data sets, available online, against which an organisation or project can benchmark its performance against the national trend. There are ten 'headline' KPIs that show current performance being achieved across the construction industry, available from the Construction Best Practice Programme. They are:

- Client satisfaction – product
- Client satisfaction – service
- Predictability – cost
- Predictability – time
- Profitability
- Productivity
- Defects
- Safety
- Construction – cost
- Construction – time.

Obvious omissions are quality, communication and conflict. The first Design Quality Indicator (DQI) was launched in July 2002 aimed at helping to measure design in relation to factors such as functionality and impact. Communication, although implicit in the indicators listed above, has yet to be addressed; however, this does not prevent an organisation or project manager from creating an indicator for communications on which internal benchmarking can be carried out. For example, it would be useful to know who the better communicators are in the temporary project team, why they perform better and thus identify factors that may help others to communicate more effectively. Another benchmark that appears to have gone unnoticed is that associated with intrapersonal communication. Reflecting on how well we dealt with a particular situation, what we did well and what we did not do so well is a form of personal benchmarking aimed at improving our performance and interpersonal skills.

Benchmarking and the use of key performance indicators as a tool for measuring progress are important elements in the desire to improve performance and hence effect change. As with any change management programme it is essential to discuss the introduction of benchmarking with employees and/or project members before implementation. There are a number of simple steps to follow:

Development

- Explain and discuss purpose of KPIs, their benefits, use and development
- Align the use of KPIs with organisational/project improvement strategies, i.e. relate them to critical success factors for the organisation/project
- Agree a timescale and process for KPI introduction, use and development
- Determine whom/or what to benchmark against (e.g. a competitor, a previous project)
- Ensure adequate staff training is in place before implementation.

Implementation

- Monitor the collection of data and the display of results/performance
- Set realistic targets and review at set intervals
- Determine current performance gaps and identify reasons for differences
- Continue to systematically review reporting of data and the display of results
- Discuss progress with participants and incorporate feedback
- Refine and modify KPIs to retain their relevance to the organisation and projects.

It is also necessary to look at the issues of safety and quality before turning our attention to the design and maintenance of the project culture.

Safety and communications

Worldwide the construction sector has a very poor safety record. A number of factors are to blame, from inadequate training, poor adherence to guidelines and legislation, unsafe working practices, etc. to trying to do too much in the time available. Various layers of legislation exist which try to prevent or reduce the incidence of accidents, but despite the threat of heavy financial fines and a custodial sentence for non-compliance, death and serious injury still occur. In our discussions with site personnel we were alarmed to find that there was a high degree of ignorance of safety legislation among site personnel and the temptation to flout even the most simple safe working practices was only too evident. There is a cultural problem to overcome and this is linked in part to the ability and determination to inform individuals of their obligations regarding safe working practices.

Through our research into site-based communication we have also noted a number of practices that affect the response to safety issues. They are:

- Lack of ownership of safety legislation
- Confused responsibility for following and enforcing safety procedures
- Failure to integrate safety into management and design practices
- Lack of response to non-compliance with safety policy
- Poor communication of safety information to site operatives.

On a number of occasions management representatives would ask site personnel to wear their safety helmet, flourescent vest or protective footwear. When messages were not delivered firmly and sanctions not enforced for those who did not comply, the individuals concerned continued to disregard safety procedures. When personnel were clearly informed of legal requirements and company policy, and were informed of the consequences of not adhering to the law, the personnel were found to change their behaviour. The message was firmly stated and the position of the company with regard to safety was reinforced. Access to the site was refused to anyone without the correct attire or training, and operatives who did not adhere to safe methods of work were removed from the site. We also found that the firm and reinforced delivery of safety procedures resulted in less time spent reminding people of safety procedures. It is necessary for site managers to:

- Issue clear instructions and state the consequences of non-compliance
- Confirm that legal requirements are not negotiable
- Ensure that non-compliance with safety procedure results in positive action.

While the CDM Regulations state that safety must be considered both at the design and management stages, during our observations of management and design team meetings none of the sites we observed discussed safety issues when proposing changes to the design or work procedures. The only health and safety issues that were discussed during progress meetings related to accidents or reportable incidents that had occurred since the previous meeting. When proposing changes, attention should be given to the method of work, programme of events and resources allocated. By giving prior consideration to how works are to be managed it is easy to ensure that practices are safe and that adequate equipment and resources are allowed for. Good, safe working practices offer the best potential for a quality product. Where contractors or sub-contractors use fewer resources and equipment than specified in their method statements and programmes, the potential for the quality of work to suffer and safety practices to be breached is increased.

- Method of work, sub-contractor programmes should be fully integrated with safety documentation
- Operatives should be fully aware of the safety procedures
- Procedures required to undertake work in a different way from that stated in method statement should be followed.

Organisations cannot ignore safety if they are to remain commercially viable. The need to collect background information on those employed is enforced by the CDM Regulations, which state that only competent contractors with good safety records should be employed. Emphasis on information exchange, ensuring health and safety procedures are communicated, and that safety is considered when designing, constructing and maintaining the building, places considerable importance on communication practices and the safety culture that they may or may not instil. To ensure that the transfer of safety information is effective, a safety culture must be developed that is receptive and reactive to health and safety issues. A convincing argument for a proactive approach to the communication of health and safety information, thus helping to improve awareness and in turn reducing accidents, has been put forward by Preece and Stocking (1999). They identified the absence of feedback, selective attention to messages, the lack of sender credibility, use of too much technical jargon, filtering of information and problems with status differences as factors that hindered the communication and hence awareness of safety legislation and practices. Building on their advice the project manager should:

(1) *Encourage feedback and the use of active listening* Managers should encourage feedback that helps to identify where information is unclear or needs further explanation. Even if people understand the information, further clarification can help to reinforce the message. Managers and designers should be encouraged to listen to operatives and encourage feedback. Time should be devoted to listening. Parties engaged in listening should give their full attention to the sender, providing non-verbal and verbal signals, which respond to the sent message.

(2) *Regulate information flow and balance repetition* Managers should ensure that the information does not overload individuals. The amount of information supplied to individuals should take account of their personal capability for information processing and the situation. Information should be relevant to the people involved and the tasks being undertaken, thus information should be regulated and differentiated between task-specific and background safety information. Phasing information can be a useful strategy. Where information is important the message can be reinforced or reiterated, using a different communication medium or alternatively using a training event. Repeatedly delivering packages using the same communication methods may become boring and monotonous; the attention levels of those listening or receiving the message may drop.

(3) *Provide user-orientated communication* Our interpretation of messages is different because our visualisation and understanding of situations is based on personal experiences, education, training, attitudes and emotions. When communicating we must recognise that others do not have the same experience and knowledge. Attempts should be made to tailor communication to that of the user, use prior interaction to determine levels of understanding and frame communication in a way that will be easier to understand. When managers tailor information to the user's levels of education, values and experiences, communication barriers are removed. Examples are: frame of reference, selective listening, sender credibility, value judgements, information overload and use of inappropriate language. When messages are transmitted users will be more inclined to ask questions, thus encouraging a greater understanding.

(4) *Provide appropriate and carefully timed messages* Information regarding safety should be delivered at an appropriate time. The initial safety meeting at the start of a project may cover all items of health and safety, but some specific issues may be forgotten when engaging in work during later stages of the project. When operatives are about to undertake a specific task the relevant safety information should be communicated to them prior to work activities commencing. Managers should avoid giving safety briefings during lunch breaks or towards the end of the day. People are often more concerned with their lunch or going home than the information being communicated. Noise and other distractions can cause selective listening, where a person sub-

consciously focuses on an event that is particularly salient to them. Unfortunately, if distractions are present the most salient information may not be the safety briefing. Some managers choose to transfer information in hotels, universities or other off-site accommodation; however, this can be expensive. Simple arrangements that consider the impact of the environment and time of day on the recipients of the safety information may be just as effective.

(5) *Improve upward and downward communication*   Managers often overestimate the amount of information that is transferred down an organisation's hierarchy. Feedback and face-to-face interaction that removes these boundaries can help to reduce barriers. Management can use employee suggestion schemes, open door policies and group workshops to facilitate decision-making. Participation in decision-making can improve upward communication and encourage ownership, as the employees are involved in the development of the company policy. If comments are requested by survey, the survey should be anonymous, encouraging open comments.

Communication of safety information

One of the most important aspects of safety training is that everyone feels that they have ownership of the safety policies and considers them to be their responsibility. All contributors have a role to play here, although it is incumbent on the project manager and the construction manager to ensure that the communication of safety information to employees is effective. A variety of methods are useful in helping to achieve this objective, being:

(1) *Induction*   Every party entering on to a construction site should be informed of safety procedures that apply to the site and the potential hazards they may encounter. Individuals should also be made aware of their working areas and areas prohibited to them.

(2) *Toolbox talks*   These are used to deliver brief safety information on site where the work is taking place. Discussions are directly relevant to the particular workgroup being addressed.

(3) *Videos*   Videos are useful for many purposes, for example, delivering safety information that is common to all parties working on site or in an organisation; helping to make employees aware of new regulations; and highlighting specific dangers, such as trench collapse.

(4) *Training days*   Where it is necessary to transfer large amounts of safety information to a large number of people, then a training day can prove very useful. These are usually delivered off site, thus removing unnecessary distractions. A number of presenters should be used and the day should be broken up with discussions, workshops and other activities which prevent people becoming bored and help to get the messages across. It is important to reinforce the information at a later date, perhaps via a toolbox talk.

(5) *Workshops*   Those present at the workshop engage in discussions, activities and in decision-making groups that aim to resolve a specific safety problem. Workshops help to develop practical skills for dealing with safe working practices and also help to encourage a sense of ownership of health and safety.

(6) *Presentations*   Presentations are a useful tool for transferring information, and a mix of presenters internal to the organisation and from without helps to keep issues pertinent and current. Audience interest must be maintained throughout, and short, frequent presentations are better than long, infrequent ones.

(7) *Tests*   The only sure way of knowing if an organisation's members understand company policy and procedures is to test them. A number of methods
of testing employee knowledge are available, for example, simple question-
and-answer sessions at the end of presentations or the use of a multi-choice
tests at the end of a video presentation. It is important to point out that getting
answers wrong does not necessarily mean that the employee is the problem, it
is just as likely to be a problem with the way in which the information was
conveyed. This feedback is essential if organisations are to have some confidence in their employees' knowledge of health and safety legislation.

(8) *Conformation slips*   Conformation slips can be used to show that an employee
understands safety information. Many people believe that when an employee
signs a form to say they understand the health and safety policy that the
organisation is relieved of its responsibility to ensure safe working practices.
Safe working practices need to be constantly reinforced.

(9) *Helmet stickers*   Many companies use Health and Safety stickers, which are
signed and dated, to show that they have received safety training. These are
particularly useful for those who supervise and manage the site. An operative
who does not have a sticker on their helmet can be asked to leave the site until
they have been inducted.

(10) *Safety manuals*   Safety manuals, company policies, CDM Regulations and
COSHH (Control of Substances Hazardous to Health) documents should be
readily accessible on the site. These are useful aids to recall safety knowledge,
although due to the amount of information contained in the documents users
should not be expected to read them from cover to cover. Instead they should
be aware of what relates specifically to their own area of work and their
responsibilities in relation to their colleagues on the site.

(11) *Method statements*   Method statements and programmes of work are
particularly useful in the management of safety. Contractors and sub-
contractors are required to produce method statements that state how specific
packages of work will be carried out safely. Method statements should make
reference to the number of people allocated, plant required, phasing and
timing of events and any procedures and regulations that are followed. Many
method statements include short-term programmes that show sequences of
events and the resources to be used. Failure of a sub-contractor to use the
resources and equipment specified, or undertake the operations as stated,
indicates that a safety or quality procedure has not been followed.

Communication and quality

Closely linked to the communication of information about health and safety is the
communication of information relating to quality standards. The quality required
for materials and workmanship will be stated in the written specification, which
will also refer to relevant standards and codes. The project manager must work
closely with the site manager (construction manager) to ensure that this information
is read and understood by those needing the information. In some respects the
issues are similar to those involved in conveying appropriate health and safety
information in that there is a constant need to reinforce the required quality standards and ensure that operatives understand what is required of them. This is
particularly pertinent on sites that involve many different trades and many different sub-contractors trying to work together, often within a limited space and a
fixed timeframe, and to the same standard. In addition to the quality control and
quality assurance procedures in place on the site, the appropriate managers must
ensure that all site personnel are familiar with, and understand, the following:

- Written specification (quality of materials and workmanship)
- Any special or unusual quality requirements
- International and national standards
- Manufacturers' instructions (storage, safety, fixing, protection, maintenance, etc.)
- Requirements of planning consent (e.g. approval of materials).

Areas that deserve special attention are:

- Design changes
- Substitution of specified products (change of supplier)
- Changes to the sequencing of work
- Changes to staged information
- Changes to personnel/sub-contractors on site.

Like health and safety, quality standards need to be discussed with those involved in realising the design and the constant reinforcement of the message can help to foster a quality culture. Developing the right attitude at the outset of different works packages has been found to be useful in the drive to improving standards of workmanship on site. This is particularly important in the current climate of using sub-sub-contracted labour where the message can too easily be lost.

Designing the project network structure

As intimated at various stages within the book, managers may have very limited control over the effectiveness of communication. Within their own organisation, be it architectural or contracting, there may be formal managerial structures that help to ensure effective communication, but at the boundary condition and especially within the temporary project structure, control becomes more tenuous, relationships and reactions less predictable and the management of communication more challenging. Of course, managers must not use this as an excuse for communication breakdown. Identification of the temporary project network's barriers to effective communication, combined with the realisation of their own limitations, will make managers better equipped to control the process and hence affect the quality of life for all those involved. But control will be limited.

Construction projects constitute a temporary arrangement of diverse and often competing groups, entering and leaving the project at different times depending upon the stage of the project, forming temporary communication networks in the process. The challenge for the project manager is to recognise and then manage the different communication networks to achieve a successful outcome, i.e. he or she acts as a facilitator of information exchange and also as a key decision-maker. Project managers must also concern themselves with how different groups relate to one another, how the groups may change over the period of a project and how the groups communicate. Thus a project manager's main concern at the outset of a project should be taken up by addressing how the diverse range of professionals will interact during the life of the project. Will they all get along? Will they all demonstrate the same level of commitment to the project goal? These and associated questions must be asked at an early stage. At the back of the project manager's mind is the realisation that life is messy and human beings rarely behave as expected (or desired). Essentially he or she needs to have a thorough understanding of group dynamics, with the ability to design the project's culture from inception and manage its instability for the duration of the project.

Selecting the right people for the project

To achieve a successful project outcome we need professionals who can, and are willing, to work together. These may not necessarily be recognised as the best in their field, but combined they will work towards a common goal. As individuals we react differently to different people. In a social setting we are free to choose our friends and avoid those we may find irritating. In a work environment we may have little choice about whom we share office space with and so we have to act professionally and try to get along with everyone equally. This also applies to projects where people are brought together because of the project, they may not take an immediate liking to one another. This will influence communications. It is important to know which individuals within an organisation will be working on a particular project (get the organisation to put it in writing and adhere to it).

- Will those selected help to realise KPIs?
- Do all organisations embrace TQM?
- Are there any potential barriers to prevent participants from communicating effectively?
- Is enough known about the organisations and the people involved to make valid decisions?
- Have organisations/groups/individuals worked together before? If so, was the arrangement successful?
- Do the participants share mutual trust?
- How likely is conflict between organisations?

Composing the communication culture

The project culture should be designed before a project gets underway with the aim of improving group co-operation and communication. Equally, members of 'periphery' groups associated with the project are dependent on the project co-ordinator listening to their requirements and concerns, incorporating them into the project knowledge base and transmitting them to the relevant parties, thus providing a 'real' chance to participate in the project.

As intimated earlier, texts on procurement routes have, arguably, missed the point. It is not the type of procurement route that is important, it is the manner in which the participants work together towards the realisation of a common goal. Selection of organisations and participants who are happy working together and who share mutual trust is the most important factor, the contract employed is a secondary concern, merely a tool to facilitate formal relationships between project participants. If communication is effective there is less likelihood of disputes arising, but when communications break down disputes will follow which may lead to legal action and which will need to be dealt with through arbitration, alternative dispute resolution (ADR) or the courts. As a response to changing requirements and to assist different parties in managing their exposure to risk, different procurement systems have been developed. Some of these may help to improve the level of integration between design and management, thus attempting to reduce the impact of fragmentation on the construction process. Design-and-build projects, where one organisation takes full responsibility for both the design and management packages, should in theory foster a more integrated approach. Dainty and Moore's (2000b) research into teams operating on design-and-build projects found that communication barriers and practices prevented the development of an integrated project culture. Their report found that the complex nature of interactions, inconsistent membership of workgroups and the physical proximity of participants presented considerable difficulties in addressing cultural boundaries. The traditional methods of operating continued in spite of contractual procedures and management systems used by the main contractor. The design team – architect, structural engineers, building services engineers and other specialist designers – had their own cultural practices and professional identities which resulted in distinct cultural interfaces that on many occasions bypassed the main contractor. Such practices caused confusion and conflict when dealing with unexpected changes.

To foster an integrative culture, management systems and practices should be established from the outset of the project. Early involvement of both management and design specialists is essential. The flow of information and communication procedures must be configured so that an integrated project culture is embedded early in the process. The project culture is not just about forms and documentation, it is about informal processes, levels of congruent understanding, perceptions, behaviour, work ethics, practices and beliefs. To develop a positive project culture, events are required at the earliest possible juncture to help foster and develop the correct business climate, which will help instil and develop integrated work. As the desired project culture emerges, further group events will be required to reinforce and maintain the dynamism.

Whether communications are effective or not will depend upon the talents of the individual(s) controlling the communication routes and their gatekeeping skills just as much as the technologies employed to facilitate the tasks. As mentioned above, the project manager acts as a node through which all communication should flow, as such it is without question the most important role in the project team. Control of information and communication routes is the key to control of the project. It follows that the project structure must be designed.

Controlling communications

Hastings (1998) suggests that complex projects require an overall communication strategy, identifying ways of reducing cultural and language barriers, identifying networks and responsibilities, introducing informal communication to meetings, structuring meetings and objectives and building relationships, in order to ensure project success. The flow of information during the project will be through the project co-ordinator. Project managers are essentially information and communication managers, collecting information from specialists and supplying others with enough information for them to make informed decisions. Participation in construction projects is influenced by the effectiveness of communication between individuals and the manner in which information is disseminated and managed within and between networks. The communication networks must be given careful consideration and designed with as much care as the building itself.

Designing the project structure

The design of the project organisational structure, the temporary project network, is an important function for the project manager. In Chapter 7 we identified the three levels of networks that are present in construction projects. These are the:

- Formal network
- Statutory network
- Informal network.

The project team is formally constituted through contractual arrangements, usually appointed individually by the client or client's representative. It is important to recognise that this formal team is influenced at various times by the contribution from individuals with no contractual link, for example the town planning officer, the building control officer, local pressure groups and the building users; each with competing values, different goals and varying cultures, themselves forming part of the statutory network. It is useful to reflect on this observation from a project manager's perspective, since all three networks will exert different demands on both the project manager's managerial and interpersonal skills.

The constitution of the first network, the formally constituted project team, is under the control of the project manager and, depending upon the individual's level of involvement, the client. It is the project manager's remit to assemble the best possible team for a particular project; that does not necessarily mean the consultants with the best credentials, but consultants who are best able to communicate with each other. The temporary project team is a social structure and the manner in which the participants interact will determine the effectiveness of their communication and the success (or otherwise) of the project. This is an important point to make, the project manager selects the designers, the structural engineers, the contractor (by way of the select tender list, for example) and some, if not all, of the sub-contractors (through nomination). There can be no hiding place if things go wrong, the initial choice of consultants will set the tone for the entire project and form the project culture.

The second network may be less easy to control since the network is determined not by the project manager but by the physical location of the site. The local authority responsible for a particular site will determine statutory contributors to the project, for example, town planners and highway engineers. However, the

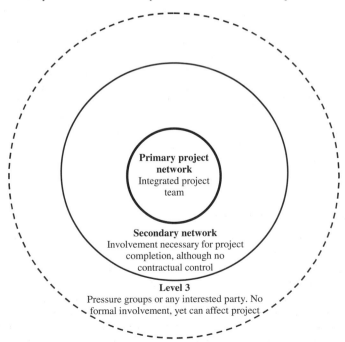

Ability to control and influence parties reduces as the network expands

Primary project network
Integrated project team

Secondary network
Involvement necessary for project completion, although no contractual control

Level 3
Pressure groups or any interested party. No formal involvement, yet can affect project

Figure 9.1   Project networks and control.

project manager should be able to anticipate the contribution of, say, the planners, based on his or her experience of the particular local authority and any particular idiosyncrasies they may have revealed on previous projects.

The third network, the informal network, is beyond the control of the project manager, all he or she can do is 'manage' the relationships that develop to the best of his or her ability. For example, the project manager cannot influence neighbours and local pressure groups, nor can the extent or timing of any contribution (usually through the planning authority) be anticipated with any degree of accuracy. Thus the project manager is forced to be reactive and sensitive to any external messages, filtering them and transmitting relevant messages to the consultants for their consideration.

Communication will take place within and between networks and it is where groups interact that needs to be effectively managed so that there is no loss in the quality of information transmitted from one group to another or from one network to another. Thus the project manager must be aware of group dynamics and responsibilities throughout the project's quite diverse stages in order to manage the process effectively. He or she must design the project's networks and hence the project's culture before the project gets under way, a proactive rather than reactive approach to project management.

Implementing project intranets and extranets

Information technologies have enormous potential as tools to facilitate effective communication within the project context. Project intranets and extranets are being used successfully, allowing easy access to project- and product-specific information. The ability to transfer information quickly and record transactions for future reference is a big asset. However, not all consultants, contractors and sub-contractors have made the move to the paperless office and hence the paperless project. Many operatives and consultants still need to see printed copies of drawings to fully understand their subtle meanings – somehow the printed version always looks different to that on the computer monitor. Again, we need to be aware of the user's ability to understand information contained in different formats.

The issue of software compatibility also deserves a brief mention, as it can be a problem if not dealt with early. Despite the manufacturers' claims as to compatibility it will be necessary to check that different CAD systems, project management and or data transfer and capture systems are compatible. Clear protocols for using ITs need to be established at the project outset and adhered to throughout the life of the project. It is worth remembering that things can go wrong just as easily as using more familiar, paper-based, systems. Any system is only as good as the people using it.

Communication breakdown

It is an obvious statement to make, but it is vital that communication breakdown, no matter how minor, is dealt with quickly and resolved to the satisfaction of all parties. Failure to do so is likely to lead to dysfunctional conflict, which will take time (better spent on more productive activities) to resolve. This means that the project manager and managers of specialist groups and teams must be vigilant and constantly monitor interaction, thus helping to ensure the project continues to run smoothly. Managers within organisations must deal with internal conflict and work with the project manager to recognise and respond to interorganisational conflict and interprofessional conflict if it arises.

Further reading

Fryer, B., Egbu, C., Ellis, R. & Gorse, C. (2003) *The Practice of Construction Management*, 4th edn. Blackwell Publishing, Oxford.

Walker, A. (2002) *Project Management in Construction*, 4th edn. Blackwell Science, Oxford.

10 Selecting appropriate communication media

To ensure that a message achieves its desired effect, it is essential that the method used to transfer information supports the communication process. The choice of media to achieve information transfer will depend on the nature of the situation and the parties involved in the communication act. Different communication media are evolving, the speed and styles used to transfer information are changing; however, the new technological methods of communicating can be wasted if the people and the processes involved are not given due consideration. This chapter looks at the media used by construction professionals and explores the media used during the design and construction phases. Traditional paper systems and IT tools are discussed before turning our attention to co-ordination of information chains.

Choice of media

Closely associated with the establishment of a project culture are the issues of managing communications and the correct choice of media. In many respects we are still concerned with the issue of control, that of designers trying to control (and maintain) their design intent and that of construction managers trying to control the works to achieve the design intent as represented in the contract documentation.

Designers work and communicate indirectly. Their creative work is expressed in the form of instructions to manufacturers, other consultants, contractors and sub-contractors, and is usually in the form of drawings and written documents collectively known as 'production information'. It follows that manufacture, design and construction rely on effective communication to achieve quality artefacts. Instructions must be clear, concise, complete, free of errors, meaningful, relevant, accurate and timely to those receiving them. At every stage in the process the ability to communicate is essential. Designs need to be explained and defended to colleagues, consultants, planners, project managers, contractors and, of course, the client. Discussion, argument, compromise, and (hopefully) agreement, are integral parts of the whole, thus communication media need very careful consideration. The selection of one medium over another is chosen to satisfy a particular set of circumstances. Yet the ability to communicate design intent from client, through inception, detailed design, tendering and then assembly on site is often taken for granted, especially the verbal and written elements. Professionals should get it right every time, but for a variety of reasons – such as mis-interpretation, omission and error – communication can be ineffective, leading to aborted work, increased costs and time over-runs. In the worst cases this can result in expensive and protracted disputes from which few, other than the legal profession, benefit.

Effort, discipline and commitment are all necessary if ideas, technical data and instructions are to be transferred to a diverse range of people effectively and efficiently. Sending all of the information to everyone involved in a particular project is now possible with electronic distribution through the Internet or intranet; however, the issue of relevance and information overload needs very careful consideration. The tools of communication are: oral, written, drawn, physical models, video, and physical gestures. All, to lesser or greater extents, will assist in the evolution of the design and the delivery of the building.

Use of communication media in construction

Given the wide range of communication media available, a good starting point would be to look at how different professionals use media. Communication skills will differ depending on the context in which they are set. Factors such as the characteristics of the individuals involved in the communication act and the environment in which communication is taking place, combined with time pressures, may determine what constitutes effective communication. However, research conducted in other business environments seems to suggest that there is a common hierarchy of communication skills that are required by the majority of organisations. A survey by Di Salvo (1980) examined 25 organisational communication studies. From the survey a set of common communication skills emerged. The most common skills identified included: listening, written communication, oral reports, motivating, interpersonal skills, information interviewing and small-group problem-solving. Earlier work ranked the communication skills most needed in order to perform effectively within a public service environment (Murray 1976). Ranked in order of importance they are:

(1) Oral communication
(2) Written communication
(3) Interpersonal skills
(4) Group leadership abilities
(5) Ability to persuade
(6) Small-group dynamics

Although oral, written and interpersonal communication skills are ranked highest, research has shown that there are significant communication skill deficiencies in these areas (Woodcock 1979). It would seem that where effective communication skills are most needed they are often poorly exercised. For example, media may be incorrectly used; the information conveyed may be incorrect, confusing to the receiver, conflicting, and/or vague; messages may be sent to the wrong person and/or conveyed to too many people; information may not be specifically addressed, thus missing its target; and there is no attempt to follow up the communication to check that the message was received and understood. Given such an observation we may wish to consider earlier recommendations that important communication within organisations should be communicated orally, or orally and backed up in writing, to avoid any confusion (Smith *et al.* 1977). The effectiveness of transmitting information to business and industrial employees indicated the most effective as combined oral and written, oral only, then written only, and then the bulletin board and grapevine as the least effective (Dahle 1953).

Unable to find any studies on the use of different communication media in construction, we undertook a modest piece of research to get a feel of the media used by construction professionals. We gathered our data through the use of a postal questionnaire, semi-structured interviews and observations of site managers' behaviour on construction sites (see Gorse 2002 for full details).

Questionnaire survey

Data were collected from construction professionals about their perception of the effectiveness of communication media within the construction context. Questionnaires were posted to 600 professionals in the UK, half of which were architects and half construction managers. A total of 162 (27 per cent) completed questionnaires were returned. Each participant was asked to rate the effectiveness of each communication medium listed. Using a Likert interval scale ranging from one to ten, professionals rated the effectiveness of communication media. A rating of ten was awarded to communication media that were considered most effective and one awarded to the communication media that were considered least effective.

Statistical tests (Friedman test) showed that the ranking across the groups were significantly different. Face-to-face communication was ranked much higher than the other communication media. The next most effective media were letter and fax (no significant difference was found between these two), both included the use of drawings to convey information. Interestingly, there was little difference between the more traditional method of delivering the information by post and electronic delivery by fax. The perceptions of the effectiveness of email had the largest variation between professionals. Qualitative feedback suggested that email was not widely available on sites at the time of the survey. The lack of experience of using this type of medium may have accounted for the large variation in perceptions.

Table 10.1   Ranking of communication media surveyed

(1) Face-to-face	Most effective communication media
(2) Letter and drawing*	
(3) Fax and drawing*	
(4) Verbal communication via a telephone	
(5) Fax without drawings*	
(6) Email with drawings*	
(7) Letter without drawings	
(8) Email without drawings	Least effective communication media

Note: * The difference between ranking of communication media was not significant in these cases.

Statistical tests (Mann–Whitney) were also used to determine whether a difference of opinion existed between architects and contractors on the effectiveness of any of the communication media. The only communication media where there was a significant difference of opinion were verbal communication over the telephone (architects rated the use of telephone higher than contractors) and electronic mail (contractors rated the use of email higher). The communication medium perceived to be the most effective by both architects and contractors was face-to-face communication.

Another issue considered was whether or not professionals perceived different communication environments to have different degrees of effectiveness. Two types of context were considered to be most effective, both of which had informal communication as part of their context, and over 50 per cent of the respondents identified informal communication as the most effective environment. Furthermore, the four categories perceived to be the most effective all involved meetings and were perceived to be more effective than using the telephone, written communication or communicating through a third party.

Table 10.2 Ranking of communication environment

(1)	Informal meetings, with action confirmed in writing	Most effective
(2)	Informal meetings	
(3)	Formal meetings with action confirmed in writing*	
(4)	Formal meetings*	
(5)	Telephone conversation confirmed in writing	
(6)	Written communication	
(7)	Telephone conversation	
(8)	Through a co-ordinator	Least effective

Note: * The difference between ranking of communication media was not significant in these cases.

All of those surveyed believed that different communication media had varying degrees of effectiveness. Face-to-face communication was perceived to be the most effective form of communication medium by architects and construction managers. Only two out of the eight communication medium were perceived by architects and construction managers to have significantly different degrees of effectiveness, being verbal over the telephone and email. These results are comparable with previous studies carried out on communication media in other fields. The meeting environment emerged as a forum that was perceived to have a higher degree of effectiveness than others identified. Informal interaction was rated higher than formal interaction.

Interviews

During the interviews the majority of the professionals claimed that other parties to the contract often provided incomplete or ambiguous information, causing them additional work and, in the worst cases, delays. It was suggested that greater effort should be made to make information clear, complete and easy to understand by all contributors to the project. Many participants felt that professionals were sometimes reluctant to seek help when they did not understand a situation, that they were worried about losing face, but that a more open and honest approach would benefit everyone in the long run. Fax messages were considered to be a quick and easy way to ask questions or alert others to potential problems. Telephone conversations were used to clarify issues, discuss and resolve problems. When information was sent by post it was considered to have a level of formality that served to have a greater impact than that sent by fax or email.

The findings suggest that more informal and fast methods of communication are more appropriate for seeking help, asking questions and solving small problems. Letters were used for communicating information that needed to be recorded (with a view to any future legal action). The formality of the letter was perceived as making its contents more permanent than other types of media and appeared to carry more weight, and thus was more likely to generate a reaction by the receiver.

Some of the comments made by the interviewees highlighted the need for open communication, participation and sensitivity to the needs of others, examples being:

For a project to be successful, parties need to communicate effectively and co-operate with each other.

Professionals need to try to see things from the other's perspective.

Information given is often vague, the instructions must be clear, stating exactly what needs to happen.

Keep communication simple, informative and to the point.

Informal communication was seen to be essential for managing and administering 'day-to-day' tasks and dealing with minor problems. However, for issues that required agreement across organisations communication needed to be more structured and hence more formal. Points made included the following:

Informal environments are more effective, communication flows more freely.

Some meetings can be both informal and formal, you need a balance between the two. Informality enables people to relax and integrate, formality ensures the problems are considered properly so that the building can progress.

Informal communication by phone and on site is very useful for solving some problems, but where responsibility for the problem is an issue you need the formality of a meeting with all parties present to agree on who will do what. If you don't pull everyone together people will try to pass on the responsibility.

Observations of site practice

Ten construction sites were visited on at least two occasions to observe construction managers communicating in their natural environment. The managers observed spent much of their time working through information in order to plan and organise work for the immediate future. Writing faxes, talking on the telephone to consultants, suppliers and sub-contractors and arranging informal and formal meetings occupied a lot of their time. Other activities revolved around communicating with personnel on site and monitoring work for compliance with specified quality standards and progress. The telephone and fax machine were used frequently to communicate problems to others, raise queries, place orders for materials and to request urgent information.

When communicating with architects, structural engineers and other consultants the managers paid more attention to the preparation of the message and spent more time considering issues prior to contacting them than they did with their sub-contractors. It seemed more common for the main contractor's site staff to make 'informal contact' with sub-contractors as soon as an issue or problem arose. When problems arose that required advice from a consultant, the construction manager did not seem to attempt to make contact with the professional with quite the same speed. More time was spent analysing and discussing the problems with other staff before contacting architects or consultants. One reason for this could be that the sub-contractors are present on site more often than the consultants, thus, they are able to develop relationships based on interpersonal communication and hence they become more familiar with one another and thus more approachable. Another reason may be that the sub-contractors are in a subservient relationship to the contractor. The status of professionals appeared to affect communication, with construction managers giving more attention to their communication with those whom they consider to have a higher status. This supports the earlier work of Gameson (1992) who found that individuals interact differently with others who are perceived to have a different level of construction experience or status.

It was evident during the observations that many problems were dealt with easily and effectively over the telephone and by fax (which allowed drawings to be exchanged). However, where problems were perceived to be complex and/or difficult to describe via a telephone conversation, problems tended to be resolved with the aid of informal site meetings. In this setting the relevant individuals could meet on site, actually observe the problem and then discuss and agree a way forward. Alternatively, designers and engineers would visit the site and inspect the work before deciding on a course of action and then issuing the appropriate instructions to the contractor. When problems could not be resolved informally, they were raised at the scheduled site progress meetings. As suspected, the observations revealed that small groups of consultants and/or sub-contractors had informal discussions and agreed a course of action before attending the formal meeting, hence helping to highlight the importance of interpersonal communication outside of meetings.

Summary of the research

Different communication media were perceived to have different degrees of effectiveness, although it was difficult to assess whether or not the media managed to communicate the message effectively. In the observations we noted a high reliance on interpersonal communication to help clarify issues. Sometimes this appeared to be because the construction manager did not understand a drawing or instruction, more often it was because vital information was missing from the contract information. Clearly, more effort needed to be made by the designers and engineers to clarify and simplify the messages being sent to site, and of course to check that the information was complete before sending it. More effort is also required by construction managers to read all of the information provided to them: some construction managers would rather pick up the telephone to ask about an item of information rather than search for it in the project documentation. Although this was perceived to be the quickest way of getting information our interviews confirmed that the designers and engineers were not too amused by such antics and expected the site manager to make more of an effort. In the work reported here this problem was resolved, but both authors have experience of serious communication breakdown because of what appeared to be a lack of effort on the part of the consultant or construction manager.

Media and their different uses

Whatever combination of media is used to convey design intent from the mind of the designer to that of the individuals doing the assembly, it must be remembered that this information has uses other than a set of instructions from which to build. Letters, reports, operating instructions, maintenance manuals, drawings, schedules and specifications may be used for one or more of the following purposes:

- *As an aid to the development of the detail design before it is finalised*   Media can be used to recall information and aid the decision-making process; thus drawings, notes and diagrams are important tools for developing design ideas.
- *As an aid to co-ordination*   During the detailed design phase information is provided by a number of different providers, from manufacturers and specialist sub-contractors, structural and services engineers, to design, etc., to aid co-ordination. As the information is developed it is integrated ensuring each component functions and fits together with other components.
- *For contract documentation*   Arguably this is the main focus of the production information, used by a variety of individuals to assemble the building. The contract document provides a record of the building's requirements.
- *As a design record*   Drawings and specifications will form the main part of the 'as built' documentation. Combined with maintenance information, operating instructions, warranties and guarantees this should be handed over to the building owner on completion. Important information for the effective opera-

tion of the building and also for reference when considering alterations and/or improvements at a future date.

- *Evidence in disputes*   Should a dispute arise during or after construction then the production information and any project documentation, e.g. letters and file notes, will be required as evidence either to support or defend a particular claim.
- *Facilities (asset) management*   As an aid to making decisions, such as space planning, maintenance, remodelling etc. during the life of the building.
- *Recycling and disposal*   As a record document to aid with the effective and safe recycling/disposal of an existing building that has exceeded its service life.

Oral communication

Oral communication skills are essential. At various stages in the life of a building project designers will have to explain their ideas and intentions verbally, usually with the use of written and graphical material. Individuals will have to communicate with one another to suit particular circumstances and phases of the job. They will need to communicate within a design office, with other designers and consultants in other offices (by telephone), with those representing legislative bodies and local interest groups, and with contractors, sub-contractors, building users and of course their clients. Empathy with others involved in the project is vital if communication is to be effective.

There are a number of different situations where verbal communication is used, often aided by drawings and sketches, which needs to be recorded in writing and distributed for information/action. Not only is it good practice to record oral communication, it is an essential requirement of quality management systems.

- *Formal meetings*   Design reviews, meetings with planning officers, and site meetings will be necessary at different stages in the project to discuss and hopefully agree a way forward. They must be planned and structured in a professional manner and recorded accurately (as minutes of the meeting) with clear points of action and timeframes in which to complete the tasks. Meetings are discussed in further detail in Chapter 12.
- *Informal site meetings/inspections*   Whether these are minuted or recorded in the job diary is a matter for the way in which organisations manage their jobs. Those using quality management schemes will be obliged to record a summary of such meetings as evidence of decisions made.
- *Design reviews*   Design reviews are an excellent tool to discuss and agree project specific issues. Again the meeting should be minuted and any decisions made 'signed off' by the client for record purposes.
- *Client presentations*   The manner in which client presentations are made will, to a certain extent, vary depending upon the client (a householder or a multinational company) and the size of the project. Thus some will involve one-to-one communication during which informal presentation skills will be most effective, others will involve large-scale presentations to a client panel, committee or even an audience and require different media and more formal presentation skills.
- *Telephone conversations*   Telephone conversations are an excellent way of solving minor queries and reporting issues quickly and cheaply. The main thrust of the conversation and the agreed action should be recorded in the designer's and construction manager's individual job diary for evidence in the event of any discrepancies or disputes arising. Again, this is consistent with good practice and in accordance with quality management systems.

Written instructions

As part of day-to-day business there is a need to record matters in writing. Written reports, letters, instructions and minutes of meetings are essential for the smooth running of projects, and will be used as evidence in the event of a dispute. Compared with oral communication all written communications should be more concise, more discreet (there is no guarantee who may read them), more accurate, and free of ambiguity. Care and dedication are required in their composition, i.e. time is required to ensure that the message contained in the text is that intended. The response and feedback will be less immediate than with verbal communication (even with email).

Reports

Chappell's (1996) advice on report writing is to first ask why you are writing one. He gives three reasons:

- To give information
- To request information
- To seek decisions or approvals.

Often a report will cover all three criteria and it is important to write a clearly structured report. Fact and opinion should be clearly separated. Reports are often supplemented with photographs, drawings, programmes and spreadsheets. They may be required for any of the following situations:

- Feasibility report, outline proposals report, design report
- Progress report to client
- Technical report (e.g. highways impact study, drainage report)
- Inspection of property (condition) report
- Defects report (interim and practical completion).

Reports are read (as are drawings) by people at a time that suits them. They may be tired, short of time, impatient or disturbed when reading them, and the originator has no control over how receivers assimilate the material contained within the report. Misinterpretation can occur and the opportunity to ask questions may be limited. Therefore it is necessary to keep facts clearly separated from opinion and to write clearly and concisely.

Letters

Letters are important for requesting and confirming action or simply to bring someone's attention to a particular issue. Do not write a report if a short letter will suffice, conversely do not write a long letter where a report would have been a better medium. Letter writing is a skill and one in decline with the use of less formal media such as faxes and email. Email in particular has been criticised because users adopt a more casual (careless) form of writing. It should be remembered that email can (and will) be used as evidence if required.

Email

Changes in communication media also resulted in tendencies to use more informal styles of expression. With email and fax, messages are often short and informal.

While such practices have increased the speed of information exchange, parties should not assume that such informality is always appropriate. When communicating with a person for the first time, a more formal style of email may show respect for the other party, thus helping to establish a positive relationship.

The speed with which people reply to emails or send them makes it a fast and often effective way to communicate; however, it also means that they are often more susceptible to mistakes (spelling, typing, grammar, etc.). When exchanging information between friends this may not be a problem. When mistakes are made in business communication it can be embarrassing, parties may make a point of the sloppiness or it may tarnish an individual's reputation. Possibly the worst scenario is that a spelling or type error leads to a mistake with the message being wrongly interpreted and acted on. Such mistakes could be costly.

Schedules

Schedules are a useful tool when describing locations in buildings where there is a repetition of information that would be too cumbersome to put on drawings. Particularly well suited to computer software spreadsheets, a schedule is a written document that lists the position of repetitive elements, such as structural columns, windows, doors, drainage inspection chambers, and room finishes. For example, rooms are given their individual code and listed on a finishes schedule which will relate room number, use and the finish to be applied to the ceiling, walls and floor.

Written specifications

Drawings, models and schedules cannot convey the whole message, so they have to be supplemented with descriptive information. On very small projects this information is often provided in the form of notes on drawings; however, for the majority of projects the descriptive information is extensive and is contained in the written specification.

Specifications are written documents that describe the requirements with which the service or product has to conform, i.e. its defined quality. It is the specification (not the drawings) that determines the quality of building construction. Like drawings, specifications do vary in their size, layout and complexity. In all but the smallest of design offices it is common for specifications to be written by someone other than the designer, thus communication between designer and specification writer is particularly important. The majority of designers are visually orientated people whose skills are best employed in the conceptual and detailed design phases, therefore few have time to be involved in the physical writing of the document: this task is usually undertaken by a technologist or construction project manager, someone with more technical and managerial skills. Specification writers require an appreciation of the designer's intention and the ability to write technical documents clearly, concisely and accurately. They also need to be able to cross-reference items without repetition. Standard formats form a useful template for designers and help to ensure a degree of consistency. In the UK the National Building Specification (NBS) is widely used because it helps to save time and is familiar to other parties to the design and assembly process.

There are two types of specifications: performance and prescriptive. A performance specification is a description of the attributes required. A prescriptive specification is a statement of the proprietary building product to be used. In practice it is common for both types of specification to be used on a project, albeit for different purposes (see Emmitt and Yeomans 2001).

Schedules of work

It is common practice in repair and alteration works to use a schedule of works. This document describes a list of work items to be done, a list that the contractor can also use for costing the work. It is common practice to append the schedule of works to the specification, but it must not be confused with the specification or for that matter, schedules (as described above).

Variation orders and instructions

These are essentially a contractual way of confirming action and/or variations to contract documentation which can be costed. One must always be careful to consider the options available before issuing instructions. Difficulties are often encountered when variation orders are issued quickly in response to a problem without the various parties fully considering the implications of the instruction.

Meeting minutes

Recording the important points discussed and agreed in the meeting minutes provides a useful record for future reference. More importantly, the minutes will record agreed action to be undertaken and a timeframe for completing such action. This is discussed in more detail in Chapter 12.

Defects list (snagging list)

Defects lists, commonly referred to as 'snagging lists', provide a list of incomplete, damaged and defective items of work that do not conform to the specified standard. The items on the list will need to be rectified before the works can be considered to be complete. They are often completed during the final stages of a works package. The list may be compiled by the construction manager to ensure, as far as is possible, that all works are complete before offering areas of works to the client's representatives for inspection. Contract administrators, clerk of the works and engineers will also compile a snagging list before accepting the works. The defects list should clearly state the following:

- Date snagging took place
- Name and affiliation of the persons present
- Nature of defective item
- Precise location of defect
- Work required to rectify the problem
- Date for completion of the work.

Following the exchange of the snagging list it may be useful to hold a meeting where the party responsible for the works identifies what action will be taken to rectify the defect. Such meetings can prevent parties wasting their time attempting to 'patch up' works when the remedial work expected is much more extensive.

Drawings

Drawings are one of the most effective ways of communicating information between all members of the building team. Because there are so many different parties to a building project the complexity, style and type of drawing may vary considerably, ranging from simple freehand sketches to explain a concept, through to complex detail drawings with a specific purpose. Designers use drawings as an aid to the development of designs and details as well as for transmitting information to others. Perhaps it is because building designers spend so much time engaged in the act of drawing that they sometimes forget that reading a drawing and understanding it fully takes quite a lot of skill and experience. This is important to remember when using drawings to communicate with the uninitiated, clients and members of public, and sometimes the people on site. Committee members may have little or no experience of reading drawings – they will need some help. Another, associated problem, is that drawings are used by many different disciplines, during both the design process and construction. What is clear to one person may be less so to another. Thus we cannot guarantee that a drawing will be clear to every viewer, the originator should constantly bear in mind the risk of confusion and err on the side of simplicity and clarity. Drawings are a means to the end for the recipient, their expressive content being strictly limited to the conveying of instructions, they are not the end product in the process (Potter 1989). It also follows that drawings should be accurate. Well-crafted, visually impressive drawings are of little use unless the information they convey is correct (see Emmitt 2002).

Notes on drawings should be legible (if hand drawn), concise, relevant and used sparingly to avoid any repetition with other written documents, such as the specification. Many design offices discourage notes on drawings in an attempt to avoid duplication and errors, preferring to use the elemental drawing system. The CI/SfB elemental system comprises:

- Location drawings (Code L)
- Assembly drawings (Code A)
- Component drawings (Code C)
- Schedules (Code S).

Models

Architectural models have long been an effective way of communicating design ideas to people who are not familiar with the special language used in construction, and who may find drawings and specifications difficult to read, for example the general public and clients. Physical models have an advantage over drawings in that they represent space and form in three dimensions much more effectively than can be achieved with two-dimensional drawings. They are a particularly useful tool for developing designs and also for testing innovative details prior to production. With the uptake of computer-based drawing packages, the generation of three-dimensional models and virtual environments, physical models have declined in popularity.

Digital information and virtual details

Powerful computers and computer software are now affordable for even the smallest design office and contracting organisation, providing the opportunity for networking and the sharing of vast quantities of information.

In its widest sense the term computer-aided design (CAD) includes any part of the design activity that is assisted by computers. Designers use CAD and associated software to model designs in three dimensions, to simulate environments, provide walk-throughs for presentation purposes and to test ideas quickly and relatively inexpensively before going to production. Until relatively recently the term 'computer-aided design' was something of a misnomer, with the majority of CAD packages being little more than a drafting tool. CAD has been, and continues to be used as an aid for the more efficient production of working drawings, making repetitive tasks much easier, quicker and less tedious: more a case of computer-aided drafting than computer-aided design. The ability to import standard details held in the organisation's database or those provided by manufacturing companies has been vastly improved. Before computers, standard details had to be traced or copied on to negatives, a time-consuming process which provided very little job satisfaction for the creative individual. CAD heralded the end of the tracer or draftsperson in the traditional sense. These roles have been redefined as CAD operators and designers with proficient computer skills are highly sought after within the industry. Computer software continues to develop at a rapid pace, making the designer's job much easier and at the same time providing the opportunity to produce better drawings more quickly and cheaply. To use software packages as a true design tool the 3D object-based modelling systems provide a more user-friendly design tool than do the 2D-based ones, which are essentially a drafting tool. With the recent development of the computer-aided virtual environments (CAVE) has come the ability to test and experiment in a 'safe' environment. This has implications for health and safety as well as the detailing of buildings which can be developed in virtual reality and 'tested' before being used on site.

Networking

With the growth of cheap, powerful computers and more compatible software packages the possibility of integrating production information and co-ordinating complex information is now easy to achieve. With digital information exchange production drawing co-ordination is quicker, cheaper and, with the right software, it is considerably easier to avoid clashes between information on drawings from different consultants. Perhaps one of the greatest benefits of digital information is the ability to network from remote locations. No longer is it necessary for design teams to share the same office space when they can be working on the same project from different locations, linked through an intranet or the Internet. The integrated services digital network (ISDN) comprises recent technological developments in such areas as fibre optics, satellite communications, broadcasting and digital transmission. Combined, they form the electronic super-highways that offer instantaneous communication with high quality visual and audio resolution; ideally suited for the transmission of architectural images. For small design organisations the potential of networking to form a larger, more experienced, network of individuals with different skills and experiences is considerable. No longer do the large architectural practices have a monopoly on the large schemes. Indeed, many of the larger design organisations have embraced the opportunity to network and have outsourced much of their work to individuals who work from remote locations, home, factory or building site, thus reducing their space requirements and saving money on office rentals (and essentially becoming a network of small organisations!). With these arrangements it is becoming increasingly more difficult for a client to differentiate between large and small organisations.

The interactive construction site

With developments in computers and digital cameras it is now possible to monitor areas of construction sites to see progress and to look at problem areas without the need to physically visit the site. Remote access via an extranet or intranet can allow consultants to view the physical development of the building and to discuss problems and agree solutions online. This relatively inexpensive use of technology has been piloted successfully on several construction sites. This new means of communicating has demonstrated savings in time for the contractor and consultants, helping both to avoid delays and to assist with the smooth flow of work on site.

Co-ordinating communication media

No matter how good the members of the design team, no matter how effective the quality control and quality management system, discrepancies, errors and omissions do occur. Such errors are frequently related to time pressures and changes made on site without adequate thought of the consequences for other information (decisions made without adequate information). Many faults in buildings can be traced back to incomplete and inaccurate information and also the inability to use the information that has been provided. Discrepancies between drawings, specifications and bills of quantities can and do lead to conflict. Some of these problems can be avoided with well-designed management systems; some arise no matter how good the managerial control. Key to providing useful information is information co-ordination.

Co-ordinated project information (CPI) is a system that categorises drawings and written information (specifications) and is used in British Standards and in the measurement of building works, the Standard Method of Measurement (SMM7). This relates directly to the classification system used in the National Building Specification (NBS). One of the conventions of co-ordinated project information is the 'common arrangement of work sections' (CAWS). This lists around 300 different classes of work according to the operatives who will do the work. This allows bills of quantities to be arranged according to CAWS and items coded on drawings, and schedules and bills of quantities can be annotated with reference back to the specification. It is not unusual to find that all participants adhere to this system in part only.

Information for building design is produced and consumed by organisations that are in business to make a profit. Organisations, regardless of size or market orientation, must give their clients (customers) confidence in the service that they provide. For those involved in construction-related activities the organisations must also satisfy their clients in the quality of the finished building. Key to this is accurate and timely information that everyone who needs to can understand.

Information transmission

Information may come from a variety of sources and will be encoded in a variety of media. A challenging and interesting job for those involved in the co-ordination of the diffuse nature of information. As noted earlier, information is required for many different purposes, for developing the design, approvals, for building and for record purposes. Regardless of the media being utilised, to be effective the information should have seven important characteristics:

(1) *Clarity and brevity*   The most effective information has clarity and is concise. This is far easier to state than to achieve because it is impossible to represent everything in an individual's mind on a drawing or in text. The skill is to convey only that which has relevance and hence value to the intended receiver. This can be a matter of knowing when to stop writing or drawing. This will help the receiver to avoid information overload and enable him or her to concentrate on the relevant information without unnecessary distraction. Items should be described once, in graphics or text, in the correct place. Repetition should be avoided.

(2) *Accuracy*   It is important to be accurate in describing requirements because confusion will lead to delay and errors on site. Use correct words to convey exact instructions, use correct grammar, units and symbols, and avoid ambiguity. Words and symbols should be used for a precise meaning and be used consistently for that meaning throughout all project documentation. Instructions should be given accurately and precisely. All documentation should be complete, do not leave out important information or leave text and drawings partially complete (with a view to sorting it out later, which is a dangerous game to play). Use a limited vocabulary of words.

(3) *Consistency*   It is important to be consistent in the use of words and symbols. Use of graphics, dimensions and annotation should be reassuringly consistent throughout the life of the project and across all contract documentation. Integrated IT packages and the use of the CI/SfB elemental drawing system can help in this regard.

(4) *Avoidance of repetition*   Repetition of information in different documents is unnecessary, is wasteful of resources and, when repeated slightly differently (which it invariably is), can lead to confusion. Repetition, whether by error or through an intention to help the reader, must be avoided within and between different media.

(5) *Redundancy*   There is always a danger that superfluous or redundant material will be included in the contract information. Careful editing and co-ordination of media from various sources should help to remove the majority of redundant material.

(6) *Checking*   Everyone involved in the production of information should strive to check and double check for compliance with current codes and standards, manufacturer's recommendations, other consultants' information and compatibility with the overall design philosophy. Common problems encountered by site personnel can be reduced significantly through a thorough checking regime before information is issued. In the constant drive for efficiency and ever-tighter deadlines for the production of information, checks have been left to the individuals producing the information. Self-checking is suspect, and subject to error simply because of the originator's over-familiarity with the material. Carefully implemented quality management systems should mitigate this bad habit.

While multiple checks of documents can be useful, some checking systems can be dangerous if not implemented and monitored systematically. Many quality systems require more than one person to check work, and sign documents to say that they have checked it. While observing communication practices on construction sites, observation of quality procedures found that where more than one person was required to sign documentation to say that they had checked the work, the actual check procedure was not followed. When multiple signatures were required on one sheet of paper, the person signing the document assumed that the other person had already checked the work or would check the work at a later date. The result was that each party would simply sign the sheet without reading the documentation or checking the work. Where just one person was responsible for checking work and signing it off, greater attention was often paid to the checking process. Research in other communication contexts has found that individuals will take fewer risks than groups. When more than one person signs a checking document, perceptions of individual responsibility are reduced. In the event of a failure it may be difficult to hold an individual accountable. The parties assume that the other party has checked the work. When problems emerge each party may attempt to blame the other. When designing multiple checking procedures, it is important that each person is aware of their specific responsibility for checking and can be held accountable for failing to check.

(7) *Timeliness* Good quality information received at the right time is valuable. Good quality information received late is, arguably, valueless (except for use in disputes!). Thus, the timing of information issued needs careful consideration in relation to the various programmes that run during the design and assembly phases. A similar argument holds when requesting information: adequate time must be allowed between the request and the response.

Checklist for selecting and using communication media

There are a number of simple questions that should be raised before selecting a particular communication medium. They are:

- Does the medium help transfer understanding?
- Are all the parties who need the information able to access it?
- Will multiple formats (levels) of information help understanding or cause confusion?
- Is the medium used to exchange ideas or is it used to convey instructions?
- Does the medium assist in providing the level of informal or formal exchange required?
- Does one format of information supersede or replace a previous format?
- Will the medium be able to be used where it is required (for example computer screens are difficult to read on site when the sun is shining or it is raining)?

Further reading

Chappel, D. (1996) *Report Writing for Architects and Project Managers*, 3rd edn. Blackwell Science, Oxford.
Emmitt, S. & Yeomans, D.T. (2001) *Specifying Buildings: A Design Management Perspective*, Butterworth-Heinemann, Oxford.

11 Managing boundary conditions

Earlier we identified the primary boundary conditions and discussed inter-boundary communication. In this chapter we expand on the earlier observations with the aim of providing advice for those charged with managing communication across cultural boundaries. We conclude the chapter by suggesting a number of practical measures for achieving effective communication.

Communicating across boundaries

The importance of boundaries and their underlying characteristics was explored earlier. For those charged with managing construction projects it is vital from a communications perspective to recognise barriers and to devise strategies to overcome them, or at least mitigate their effect on harmonious relationships and communications. The cultural and operational differences between organisations and the terms and conditions of the construction contracts used will present barriers: some will be subtle and rarely noticed, others will be obvious and attention-seeking. The following, by their very presence, create boundary conditions that must be communicated across or around:

- Organisations
- Contracts
- Projects
- Construction phases
- Professional groups
- Interest/user groups.

When people work within groups on specific aspects they become familiar with the issues involved and familiar with others working on the project. Over a period of time those working closely together develop communication techniques and behaviours that allow them to communicate ideas and information quickly and efficiently, i.e. breaking down barriers. Such behaviour also allows the majority of problems to be addressed before they get out of hand and result in a dispute. Some of the communication procedures used by groups will be formal and clearly outlined in project documentation. However, many of the communication routes will emerge and change as the project develops, the majority of which will be informal.

Formal and informal communication practices will vary between organisations. As a general rule, people are naturally defensive when communicating with people in other organisations or departments. Communicators must also be aware of the fact that other organisations will operate in a different manner, thus we must assume that others may not have access to the same information or have the same level of knowledge as ourselves. Failure to recognise interboundary differences is likely to lead to misunderstanding and may lead to problems. When operating across boundaries it is imperative that we seek to identify the level of cultural similarities, common understanding and common information sources that can be routinely shared between parties. From this we can start to build up a relationship, while at the same time discovering where difficulties and problems are most likely to occur. Knowing the strengths and limitations of communication within a relationship can help to significantly improve the efficacy of interboundary communications. For example, when working with personnel in other organisations it is sometimes necessary to adopt their management systems and/or work under their particular terms and conditions to aid co-ordination and co-operation through the use of the same operating language. In essence, we must be prepared to exert a certain amount of effort in order to understand other organisations that are party to the contract, and to understand their particular needs and goals, hence the need to identify and then manage boundaries.

Boundaries will exist regardless of the procurement route employed. Individuals carry out tasks on behalf of their particular organisation, yet work within their own value system. So it is not uncommon to find that where organisations have worked well together previously, a change of personnel can have a major effect on communications. Given the long duration of many construction contracts, changes of personnel are inevitable and time must be set aside to allow the participants to establish a good working relationship with the newcomers. Our own research suggests that this is rarely considered by the project manager and new members are expected to function effectively immediately – which, of course, they are unable to do. While the newcomer is expending energy on familiarisation with the project and its various protocols, he or she will not be working efficiently, thus creating a weak link in the communication network for a short period. The supply chain can be no stronger than its weakest link, so any failure in connection – through defect of any member of the chain, or of connection between the links – will result in two disconnected pieces of the chain. Messages will take longer to understand and there may be a delay in receiving replies to requests for information. Furthermore, the potential for errors is increased significantly as a direct result of ineffective communication. Contact, and hence communication, throughout must be absolute. Yet perfect contact is not a realistic possibility. Therefore, project managers and managers of organisations must allow a period of socialisation for newcomers to the project. Other members of the project team need to understand that the newcomer will take time to become familiar with their particular idiosyncrasies.

It is also important to remember that communication is a two-way process. We are not dealing with a hierarchical communication route where information is passed down the chain, instead at all of these major interfaces there is a constant exchange of information as individuals ask questions and provide additional information in response to such requests. It is at these interfaces, the organisational boundaries, that individuals will take on a gatekeeping function, having a positive or negative effect on the messages they transmit within and without their organisation. Project managers must identify and monitor boundaries with a view to helping to maintain their dynamics (if working well) or alternatively to take some form of action to redress the balance (where there is room for improvement). It follows that the system(s) set up to manage the project must include adequate feedback mechanisms. It also follows that the project manager must be able to trust the feedback received from different sources. This is a little easier to do when the project manager knows the project participants from experience on earlier projects.

This is about maintaining:

- Information flow
- Interpersonal contact with participants (e.g. telephone, meetings)

- Management structures, i.e. formal and informal communication routes
- Feedback from all participants;

with a view to ensuring:

- Groups are working towards their individual and group goals
- The quality and quantity of resources are appropriate to the task (especially when changes have been made to the design and or programme).

What we are doing here is arguing for greater involvement by the project manager, essentially a 'hands on' approach to the management of projects. Following this argument we need to design a management structure that enables the project manager to intervene if necessary. By that, we mean that he or she must have the authority (by way of both the contract and personal status) to control the interaction at boundaries. What happens within specific groups is the responsibility of individual organisations; however, the way in which the groups interact or network is the responsibility of the project manager. These networks must be mapped, designed and maintained throughout the duration of the project (and beyond to enable post-project evaluation and feedback). The project manager is the facilitator of project interaction.

Organisational gatekeepers

It is at the boundary where the role of the gatekeeper needs particular attention. Who are the main individuals who block, filter, assimilate, adjust and transmit messages within organisations party to the project? Obvious candidates are the job architect, the contract administrator and the construction manager, but look deeper and these often turn out not to be the major gatekeepers. For example, when we undertook research within an architectural practice to try to identify gatekeeping actions by individuals we found that the job architect did operate gatekeeping functions, but the main gatekeeper was the senior partner in the firm. All information (letters, drawings, schedules, etc.) went to the senior partner and often to a technical partner before they got to the job architect. The information had been filtered and added to en route, major problems were intercepted by the partners and (in consultation with the job architect) dealt with by them. This was a trait repeated in engineers' offices and in contractors' offices. The gatekeeping role is present regardless of procurement route, including supply chain management and partnering agreements, although their particular function may vary. For example, in the supply chain the gatekeepers should be clearly identifiable as the links between organisations, while in other arrangements the gatekeepers may be less obvious. Gatekeepers can have a positive or a negative effect on the intergroup interaction and hence will colour the effectiveness of the project to a lesser or greater extent. Clearly it is in the interest of the project manager to identify the gatekeepers and put measures in place to enhance or mitigate their influence, i.e. their influence may need to be moderated. These major boundary conditions and the associated gatekeeping function are explored further below.

Boundary condition 1: client–designer interface

Our first boundary condition is often referred to as the briefing stage, where client and designer interact to establish the design brief, a crucial document that contains the project parameters. But this boundary also encompasses the relationship with the client before a contract is entered into, via marketing, the relationship with the client during the project's contractual period and any future relationship once the project is complete. Each area requires different communication and interpersonal skills of those operating at this boundary.

The courting ritual

Clients, whatever their level of building experience, are often confused by the number of professional firms, with differing backgrounds, all (seemingly) offering to look after clients' best interests and of course charge a fee for doing so. Furthermore, the range of procurement routes and their ensuing contractual relationships often cause further confusion. Clearly, a professional adviser is required to help guide the client through the complexity. The seeking of a partner by both client and architect is akin to a courting ritual with both giving out messages to attract the other – success is dependent on effective communication. The communication of an organisation's professional services to potential and existing clients is known as promotional activity, or marketing. It is a form of carefully controlled communication. As with all communication, this is not a one-way process; clients, especially the more experienced, actively seek out information about professionals that may be of use to them. Hopefully, the outcome of the courting ritual will be individuals that work well together, communicate effectively and efficiently, grow to trust each other and progress to new projects as a real team.

Before exploring the relationship between the client and the designer that develops during a project's life cycle, it is important to look at the different types of clients, their particular organisational and commercial *modus operandi* and their particular needs. There are a number of definitions that describe differences in the levels of the client's construction experience. For example, Higgin and Jessop (1965) described clients as either 'sophisticated' (previous experience of construction) or 'naive'(no previous experience), while other researchers have used other terms to describe the same thing. What is more important than labelling, is to recognise that the client's level of experience of construction will affect the way in which they interact with professional advisers. Experienced clients have a better understanding of the construction process and when they interact with construction professionals they evaluate and question proposals and suggestions more than inexperienced clients. Inexperienced clients have to first learn and understand the construction process before they are able to evaluate and question suggestions made by construction professionals (i.e. they need to learn a new language). Some clients are prepared to do this, others employ a representative to help them in this regard.

Client organisations have their own specialist area of expertise and this must be understood by the individual taking and developing the brief. Some clients see buildings as an integral part of their business needs, they have a continuous supply of buildings to enable business activities to proceed. Invariably the designers come from a different background to their client. Synergy is vital for ensuring effective communication. Therefore background research to understand the client's business (the organisational culture), organisational needs, short- and long-term goals and the aspirations of the main individuals in the organisation commissioning the building is an important starting point.

Briefing: setting the agenda

The brief is a written document (sometimes supported with sketches) that aims to capture the client's requirements in writing. This document is then used by the designers and consultants to develop the conceptual and detailed design to meet the requirements as stated. A good brief should contain the client's objectives, the project timescale, the cost limit and an indication of the client's expectations of the finished quality of the building. The briefing process and the ensuing brief are critical to the effective development and delivery of the finished building, and empathy between brief-taker and client is crucial. The resulting brief must be able to communicate the client's requirements to individuals who were not party to the briefing process and who may never actually meet the client. Any queries will have to be channelled through the contract administrator. It follows that in situations where more than one professional is involved in the briefing process it is essential to establish exactly who is in charge in order to clarify responsibilities and reduce confusion.

Client expectations place considerable pressure on the professional advisers to seek feasible and economic solutions to the client brief. For major projects, a project manager or client's representative will normally assist the client with the selection of the project participants and procurement route. The three primary factors that are continually cited as those which are most important to clients are, ensuring the project is completed on time, within the client's budget and to the quality required. These goals will be achieved only if the briefing process is carried out in an orga- nised and systematic manner. There are two different types of brief-taker: the designer and an intermediary. The first involves communication within the office (intrafirm organisation), the second involves someone from outside the designer's office, and hence we are concerned with interfirm communication.

Briefing via the designer

It may be an obvious observation, but the individual or individuals charged with developing the design proposals should be involved in the brief-taking exercise. Written briefing documents do not, and cannot, convey the more subtle messages expressed by the client during the interpersonal exchanges in briefing meetings. Some of these may well turn out to be important points that with the benefit of hindsight should have been included. From a designer's perspective, the ability to see for oneself how the client expresses his or her requirements, the level of enthusiasm shown for certain areas and the response to questioning is paramount to producing a high quality design. Senior partners are the normal contact with clients and it is they who often take the client's requirements and develop the brief before handing it over to the designers in the office. This indirect communication is liable to misinterpretation and relies on good communication between client, partner and designer, in particular the partner's ability to pass on the client's requirements in the form of a written brief and the accompanying verbal expla- nation. Needless to say, some partners are better at this than others. In our experience it is better that the designer is involved in the process, i.e. attends the briefing meetings, to understand the subtle messages rarely captured in a briefing document, and thus saving time and difficulties with misinterpretation in the long run. However, in comparison with the brief being taken by an intermediary, the designer may well be producing conceptual designs and diagrammatic repre- sentations to discuss with the client as the brief develops – a skill that others do not possess. Thus the brief may be more comprehensive and the design process will have started earlier than if the brief had been developed by an intermediary.

Briefing via an intermediary

It is not unusual for the client's brief to be developed by an intermediary and communicated to the designer third-hand. Project managers often take on the briefing role before communicating the client's requirements to the appropriate consultants. In such situations the intermediary takes on a gatekeeping role. Although one may argue that this is very similar to the role of the senior partner discussed above, in the case of an independent project manager the individual lies outside the design organisation, thus informal communication is less likely. This places additional pressure on the project manager to make the written brief clear and unambiguous and understandable with little or no interpersonal discussion. Intermediaries should try to involve the designer(s) in the briefing meetings as early as possible, thus helping to ease communications and remove the need for transmitting important information through a third party; with the danger of losing some of the more subtle messages (see Figure 11.1).

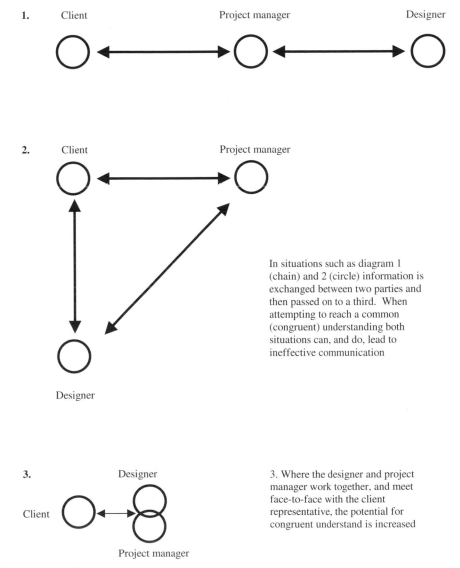

1.   Client   Project manager   Designer

2.   Client   Project manager

In situations such as diagram 1 (chain) and 2 (circle) information is exchanged between two parties and then passed on to a third. When attempting to reach a common (congruent) understanding both situations can, and do, lead to ineffective communication

Designer

3.   Designer

Client

Project manager

3. Where the designer and project manager work together, and meet face-to-face with the client representative, the potential for congruent understand is increased

Figure 11.1   Elimination of communication chains and integration of management and design functions.

What we are arguing for here is the importance of the face-to-face meeting. The problem is that an experienced project manager or experienced designer may not necessarily be good at discussing and expressing the client's requirements. What is essential is an experienced brief-taker, someone able to extract the salient points and communicate them to others. In an ideal world the brief should be developed with the designer and project manager present so that both design and production parameters can be discussed with equal weighting.

The brief-taker, regardless of background, will exhibit gatekeeping characteristics. Obvious characteristics are the reluctance to give the designers all the information (e.g. fee agreements, commercial information, etc.), more subtle ones would involve the omission of some of the client's requirements, adding their own desires to the brief when transmitting it to the designers (this is often done informally). Adding their desires to the brief may be a positive characteristic of the gatekeeping role, adding valuable knowledge to the process. Gatekeeping behaviour may well be unavoidable, but if two people with different skills are involved in the process, along with the client, then the more negative aspects of gatekeeping behaviour may be mitigated to an acceptable level and the positive characteristics enhanced.

Essential characteristics of the brief-taker are:

- A good listener
- Ability to explore sensitive issues
- Ability to record client requirements succinctly, yet without losing the spirit of the discussion
- Ability to communicate requirements to others
- Willingness to separate the client's requirements from those of the brief-taker.

Client involvement during the contract

The client or the client's representative will have an involvement in the project during the construction phase. Some clients are content to stay away from the site and the running of the project as long as the progress reports and payments continue to progress as planned. Other clients want to be involved and have the opportunity to participate in 'their' project. They will be keen to be involved in design reviews and to attend site progress meetings.

Reporting back to clients is an important function of the contract administrator, regardless of how actively involved the client is. These reporting skills are fundamentally different from the skills needed to administer and manage the construction project. More emphasis needs to be placed on diplomacy, sensitivity and tact. This helps to explain why senior members of an organisation tend to communicate with the client, not the contract administrator (often much to the client's irritation). This is particularly pertinent when problems arise and issues need to be discussed with the client.

Another area that demands clear and efficient communication concerns changes to the design during the construction period. Changes usually result in additional, not less, work and additional cost and must be handled within the spirit of the contract. Some changes are unavoidable because of unforeseen problems encountered on site, others requested by clients, designers and contractors could usually have been avoided by better communication and agreement prior to the start of the contract. Maintaining clear communication channels and a constant reminder of contractual responsibilities is a necessary function of the contract administrator during the contract. So, too, is the need to identify the exact reason for (or source) the change, the implications to cost (whole-life cost), programme and other factors such as environmental impact, constructability, etc., together with the cost of the redesign and the associated task of revising the drawings and reallocating resources. Once this has been done, approval from the client can be sought.

Designer–consultant interface

Closely associated with the first boundary condition is the boundary between the lead consultant, the brief-taker, and other consultants. This interaction may happen within a design office if it is a multi-disciplinary organisation, and/or may occur between different individuals in different organisations. Although the exact relationships will be coloured by the procurement route adopted, the exchange of information will be through a mixture of intrafirm and interfirm communication.

Intrafirm communication

The brief-taker will be working with other designers in the design office in a joint effort to realise the client's brief. The level of empathy between individuals in the same design office should be high and communication relatively informal and efficient. All members of the office should be working within the same organisational culture towards a common goal.

Interfirm communication

The brief-taker will also need to consult other consultants in different organisations (or different divisions in large organisations). Structural engineers and mechanical and electrical engineers are obvious examples, but many other consultants may also be involved depending on the nature and complexity of the project. The level of empathy between individuals in different offices is likely to be less than with intrafirm working and communication is likely to be more formal. Individuals in different offices will be working within different organisational cultures and may not necessarily share the same goals. A common example would be when individuals are working on more than one project at a time. As different projects will be more demanding at different times it is unlikely that the level of priority accorded to a particular project by those people or organisations working on it, will be the same for all involved at any one time, so the level of urgency and commitment may well vary.

There is a tendency for consultants to work with those organisations that are familiar to them – communications tend to be less defensive, more open and hence easier – and because of previous experience the level of trust is higher than when engaging with unfamiliar organisations. Designers tend to have a preference for certain structural and mechanical and electrical engineers because they know that they can communicate effectively with them and trust them to pursue similar aims and objectives in trying to realise the brief. They will know what to expect when things start to go wrong and are able to work using informal communication routes to resolve difficulties before they get out of hand.

Designer–manufacturer interface

As the conceptual design is developed and designers start to address detailed matters there will be a relatively high degree of interaction with manufacturers' technical representatives. Designers will be looking for help with technical detailing and specification writing, especially where the products and or manufacturers are not familiar to them. Their main concern will be to realise the detailed design within a tight timeframe, thus quick and reliable responses to requests for information are vital to the specifier. The manufacturer's representative will be trying to 'get the specification', i.e. help the designer to confirm their employer's products in the specification – hence ensuring sales via subsequent orders from the main contractor and sub-contractors.

Research has shown that the majority of designers prefer to stick to manufacturers and products that they have used previously, simply because they know how the manufacturers and their products are likely to perform (Emmitt 1997, Emmitt & Yeomans 2001). Because of this familiarity the designer can, in many situations, actually specify a manufacturer's products without communicating with them, drawing from their individual or organisation's file of favourite products. Interaction with the manufacturer, to order the materials, etc., is usually carried out by the contractor (see below). In situations where the product or manufacturer is not known to the specifier then communication will be necessary so that the specifier can reduce his or her level of uncertainty about the product, i.e. find out more about it, before deciding whether or not to use it.

Boundary condition 2: designer–contractor interface

Our second boundary condition is often perceived as the most confrontational, that between designer and contractor. The link between building site and design office is important, but one often fraught with difficulties. Despite (or because of) the enormous volume of information provided to the contractor the effective use of interpersonal communication skills is essential during this phase if the contract is to run smoothly.

Early involvement

There is a very powerful argument for involving the contractor and known specialist suppliers early in the design process. This assumes of course that these parties can be decided upon early in the life of the building project (i.e. during the briefing and conceptual design stages). When specialist subcontractors are brought into the early discussions it usually results in their products and/or services being incorporated into the design, thus resulting in the novation of the subcontractor to the main contractor. While this may benefit the design and communications between designer, contractor and sub-contractor, the downside is that it may reduce the bargaining power of the main contractor with regard to the price of the subcontracted works. Early involvement of specialist sub-contractors during the design stage can help to:

- Embed the design concept in the minds of sub-contractors
- Foster a more integrated approach to design and management
- Eliminate design problems often encountered by the specialist sub-contractor
- Improve constructability
- Incorporate disassembly, recycling and recycling strategies into the conceptual design
- Increase the availability of detail design information early in the development of the design
- Improve the function and fit between components

- Increase the level of understanding between specialists
- Engender a sense of 'ownership' of the project
- Develop communication channels (and hence trust) early in the process.

Early involvement of contractors and sub-contractors in the design process should be encouraged to develop greater integration and understanding, improve communications and improve the incorporation of knowledge. To do so necessitates greater attention to the design of the temporary project networks and associated procurement process. Experienced clients who engage regularly in construction works are entering into agreements with designers, contractors and sub-contractors as a partnership. Rather than putting works out to open tender, a few selected designers, contractors and sub-contractors are invited to negotiate to agree a contract sum. With the same group of organisations engaging in projects, business relationships develop and the potential for a more integrated approach to construction may be realised, with the supply chain becoming stronger with each new project. Where strong relationships develop quickly, informal and efficient communication practices can emerge and so any problems encountered with services or a product's ability to function can be resolved relatively easily and without recourse to defensive communication. It is, however, imperative that some degree of competition is encouraged to stop the favoured few from becoming complacent and to keep the service competitive.

Communication during the construction phase

Once the contractual agreement has been signed a trilateral communication network is established between the owner, contract administrator and contractor. The contractor successively establishes communication links with the various sub-contractors. Client requests and comments are channelled via the contract administrator – this may be an architect, project manager or contractor depending on the type of contract employed. All parties have reciprocal legal obligations to the client, which are dependent on the terms and conditions of the contract.

In traditional contracts both the architect and the main contractor have a contractual agreement with the client. The rules of engagement between the architect and the main contractor are set out in the terms and conditions of each contract that exists between the client and architect and the client and main contractor. In traditional contracts, the architect often takes on the role of contract administrator and with it the authority to give instructions to the main contractor. Thus, when problems emerge, the main contractor will look to the architect for a decision and instruction of what to do. In this position the architect is vested with a certain level of management power over the contractor. In package deals, such as design-and-build, it is less common for the architect to have a contract with the client. The client will usually have a contract with the main contractor, the main contractor then appoints the architect. In this situation the architect has less control over the design because it is the contractor who will make the decisions and who will have the power over the consultants. Obviously, these different relationships affect communications between the parties to the contract.

Communicating design during construction

The flow of both design and construction information during the construction phase is channelled primarily through the contract administrator and the construction manager. Members of peripheral groups associated with the construction process are dependent on these intermediaries for their information. Both occupy powerful positions when controlling information and communication. While the relative power and responsibility of these professionals will vary depending on the terms of the contract, both parties must work together to ensure that the specialist knowledge is exploited to the full. Successful and efficient building projects depend on the highest degree of co-operation, at every stage and between every level, and none more so than between the two functional partners: the architect and contractor (Calvert *et al.* 1995). The importance of the interface between management and design was highlighted in a case study by Emmitt (1999) concluding that the architect and contractor should work as closely as possible to reduce complicated communication routes. Emmitt (1999) put forward the view that there is a need for architectural practices to employ construction managers to manage building packages. This removes the organisation boundary between elements of management and design, enabling a closer relationship and better communications that may benefit those involved, including the client.

The vast majority of the information exchanged is compiled for specific functions and requires specialist knowledge to understand it. Very few, if any, of the parties involved in the construction process understand all of the information transferred and so discussions between management and design professionals are essential if individual aspects of the construction are to be successfully integrated together. The construction manager needs to quickly understand the drawings and supporting information so that he or she can order the necessary materials, organise appropriate labour and produce the building in accordance with the contract programme. Good communication and co-ordination between the construction manager, consultants, sub-contractors and suppliers is essential if quality and time parameters are to be met. It is here that the construction manger becomes the centre of information distribution, responsible for ensuring that the appropriate information gets to the various sub-contractors and tradespeople. Thus information flow is not just a concern of the architect and project manager, it is also central to a well-managed construction site.

Communication difficulties often occur during this phase because it is here that the level of information available to all parties reaches its peak. Unfortunately, there is a correlation between an increase in the amount of information available and increased levels of conflict (Huseman *et al.* 1977) and therefore managers should anticipate some degree of conflict. As information is received from structural engineers, architects, mechanical engineers and other consultants, discrepancies between drawings should be expected, and checks should be made to find where instructions are incompatible. Any problems must be reported to the contract administrator and (short) meetings should be held with the aim of quickly resolving any differences. This period is obviously stressful because time is limited and people may have to rework designs and reschedule works packages in a very short timeframe. The communication practices employed during meetings should aim to both resolve technical problems and ensure that the relationships between parties remain stable and functional. Failure to maintain relationships during stressful situations can result in major disputes.

Contractor–client relationship

The extent to which the main contractor and the client will interact will depend to a large extent on the contract employed. In contractor-led arrangements the interaction will be high; however, in design- or management-led contracts the formal level of contact tends to be restricted to the site progress meetings. Where contractors are contractually obliged to communicate with the client via the contract administrator it is common for the contractor to establish informal communication routes, thus bypassing the contractual gatekeeper. In some instances such relationships are quite innocent attempts to make sure the client knows what is going on; however, often these informal channels are used to try to change things to the contractor's advantage. For example, pressure on clients to change specified products and/or alter details to suit the contractor are common informal requests that would have been rejected out of hand by the contract administrator. The contractor knows that once the client becomes involved the contract administrator must deal with the request and communicate any decisions via the client. From the designer's and contract administrator's perspective, clients must be informed about the dangers of such interaction. From the contractor's perspective it is one way of getting around awkward consultants.

Site-based progress meetings

Informal and formal meetings are used to transfer information between the design-orientated professionals and those charged with assembling the building. These meetings provide a central forum where different organisations can interface to ensure that the components of the building fit together as detailed and subsequently perform as intended. Face-to-face meetings, when managed correctly, help to develop relationships and overcome problems posed by organisational boundaries. As argued earlier, group meetings should play an important part in the construction process. The meeting is a management tool aimed at ensuring the project develops in accordance with the objectives set out in the brief. The site-based progress meeting is used to review the progress of the works, to consult others, to make decisions and to record and disseminate such decisions to the various participants. It is also used to co-ordinate management and design activities; one of the main aims is that the building meets with the client's requirements in terms of time, cost, function and quality, quality being determined by specification, standards, contracts and, where not described in such documentation, negotiation. These and associated issues are explored in Chapter 12.

Boundary condition 3: construction manager–site operative interface

As the majority of work undertaken on construction sites is carried out by sub-contractors the relationship between sub-contractors and other parties should be a major concern. However, for many designers and clients the relationship between the contractor and the sub-contractors or tradespeople is of little concern, until things start to go wrong. After all, the client communicates with the contract administrator who is responsible for dealing with these issues. Designers should, arguably, be more concerned with the organisations doing the work, although they may not necessarily have any control over this under many forms of contract. The relationship between the main contractor and the sub-contractors (and sub-sub-contractors) who are employed to undertake the work is clearly important and an area that deserves more attention than it usually receives. It is here that the quality of the workmanship, trust and commitment are evident in the finished product. It is here also, that communication routes can become very convoluted and responsibilities confused as work is sub-contracted numerous times down a chain for which few really want to take responsibility. This is a real concern for the health and safety of individuals on site, and clear monitoring and control of sub- and sub-sub-contractors is an inherent part of the site's health and safety procedures.

Problems between sub-contractors and contractors can emerge when relationships are temporary and there is no prospect of further work. There is little incentive to build relationships. It is a very different picture when sub-contractors and contractors know that there is a possibility of repeat work since there is a strong incentive to do a good job. In such situations the sub-contractor has an interest in building relationships. As sub-contractors and contractors work together over a period of time business relationships develop that enable them to work more closely. Many contractors now have a directory of sub-contractors that they have worked with which includes post-project analysis of sub-contractors' performance. Thus, sub-contractors who have repeatedly worked well with the contractor have a better chance of securing work in the future (akin to the supply chain ethos). With the emphasis on supply chains and the nomination and novation of reputable subcontractors, the relationship between contractors and suppliers has the potential to improve. With increased effort to build relationships through effective communication the supply chain should become stronger and produce better results. Perhaps, from a communication perspective, the biggest danger here is with complacency. Individuals can become too familiar with others and start to take things for granted, i.e. get lazy and not communicate relevant information at the appropriate juncture.

Contractor–manufacturer interface

The contractor's relationship with manufacturers and suppliers is different to that of the designer. The designers and engineers are primarily concerned with the technical and aesthetic characteristics of the product. The contractor will be more concerned with initial cost, availability and delivery of the product to suit the construction programme in addition to prompt technical back-up on site. Furthermore, it is highly likely that the contractor will be dealing with different individuals to those who helped the designer with the specification, thus any interpersonal communication and development of relationships between designer and manufacturer may be of little use to the contractor's buying department. This means that new communication routes need to be established directly with the manufacturer or indirectly through a builders' merchant. Further communication routes may be established during the course of the project as technical problems develop on site and have to be resolved between detailer, contractor and manufacturer.

Where proprietary specifications have been used the contractor will be contractually obliged to use the specified product. However, this does not stop the contractor from requesting changes to the specification with the aim of using manufacturers and/or products that are more familiar and on which a higher level of discount may be available. When performance specifications are used the final choice of product lies with the contractor, not the designer. This allows the contractor a certain amount of latitude in choosing a product that matches the performance parameters. Again, the tendency here is for the contractor to use familiar manufacturers and the established communication routes/relationships, thus reducing uncertainty.

Practical measures

A number of very simple but effective measures should be considered at the project outset:

- First, identify the main and secondary boundaries
- Identify the participants
- Identify formal communication routes

- Identify informal communication routes
- Identify gatekeepers.

During the life of the contract:

- Constantly monitor boundaries, participants and communication routes for changes
- Try to simplify communication routes during the life of the project
- Encourage all participants to participate and to share information with others
- Ensure all communications go via the contract administrator
- Discuss all proposed design changes and ensure the implications with regard to constructability, cost and time are fully understood before making any changes
- Allow for a period of socialisation when there are changes in personnel.

When working across recognised boundaries:

- Allow extra time when communicating across boundaries
- Attempt to understand and appreciate cultural differences
- Expect management systems to be different – ensure all relevant information is transferred when using different systems
- Do not allow communications to break down – ensure positive relational communication is maintained, especially during stressful situations
- Develop a good rapport with staff in the other organisation
- Do not allow communicators to become complacent (and hence fail to communicate)
- Highlight potential problems at the earliest opportunity and work with the organisation to resolve problems
- When communicating important information use multiple forms of communication media to aid understanding
- Ensure people are aware of when you want to them to take action, remind them;
- Ask questions to avoid misunderstanding
- Attempt to work with people who are trusted by you and who share mutual understanding.

Further reading

Blyth, A. & Worthington, J. (2001) *Managing the Brief for Better Design*, Spon Press, London.

12 Managing meetings

Potentially one of the most effective mechanisms for ensuring productive communication is the forum of the meeting. In construction there are many types of meetings and although the meetings have different functions, the main purpose is the same, namely to review, discuss unresolved issues, make decisions, record and then communicate those decisions to those who may need the information. In this chapter we analyse the meeting forum as an aid to more effective communication and consider the problems of ineffective meetings. We also look at the issue of decisions made outside the formal meetings, i.e. in discussions prior to the commencement of, and those immediately after, the meeting, because it is here that many of the most important decisions are made.

Meetings as an aid to effective communication

No matter what our business activity, there is no escape from the ubiquitous meeting, and we all complain of being involved in far too many. Meetings are fundamental to construction; they help develop the relationships that form the group's social system, which is then used to support the technical decision-making processes (Higgin & Jessop, 1965). Meetings are a very powerful aid to effective communications and managed professionally they are key to the effective transfer of information between project participants while also providing a mechanism through which relationships are formed and socialised. Unfortunately they can consume a vast amount of time and resources. For example, construction managers tend to spend around 15 per cent of their time in meetings (Watson 2000) while other professionals have reported spending between 25 and 75 per cent on this activity (Rasberry & Lindsay 1993). Because of their importance they need to be managed professionally. Get it wrong and meetings are wasteful of resources and can become a hostile environment in which communication becomes ineffective. Organisations use meetings to exchange information, discuss challenges and opportunities, generate ideas and make informed decisions. Meetings can be classified as one of two types, either 'internal' or 'external' to the organisation.

(1) Internal meetings are limited to the organisation's members only (or in large organisations to a particular division's members). In this familiar environment it is possible to be relatively informal and trust the others present at the meeting. Discussions tend to be relatively open with shared objectives. Examples of such meetings are staff meetings and design team meetings.
(2) External meetings include the presence of members from other, possibly competing, organisations or divisions. In this environment people are expected to act in a more formal manner and will, naturally, be less trusting of others at the meeting. Discussions tend to be relatively guarded and objectives may well vary between participants. Examples of external meetings include pro-

duction meetings, site progress meetings and sub-contractor meetings and meetings at which the client is present.

Function of meetings

Meetings are held because people who have different jobs have to co-operate to accomplish tasks (Drucker 1995). An individual's knowledge and experience are often insufficient on their own, thus the knowledge and experience of several individuals must be brought together and the meeting provides a convenient forum in which to exchange ideas and agree a plan of action. So the prime function of the meeting is a co-operative forum to agree action, i.e. it is a tool to get things done. In addition to facilitating the exchange of information and decision-making activities, a review of management research literature found that meetings are also used to:

- *Control* Follow-up information, manage scarce resources, issues and create deadlines. Ensure that the managers stay in control of the tasks
- *Appraise* Meetings are also used to appraise staff, assessing management ability and how well people participate in meetings
- *Bond* Meetings fulfil a fundamental human need to communicate and bond. They create a sense of belonging and reflect the collective or cultural values of the organisation.

Meetings should not be considered as isolated events where decisions are made; instead they need to be seen in the wider context. This includes the incremental cycle of social interaction that is used to share and process information, make and confirm decisions and develop and maintain relationships.

Aims and objectives

There is little point in holding a meeting for the sake of it; equally there is little point in holding a meeting without clear aims and objectives. As a forum for sharing information and making informed decisions, meetings are necessary, but they are expensive. Thus meetings should be used sparingly and managed professionally. Two questions that must be asked before arranging a meeting are:

- What is the purpose of the meeting?
- Is a meeting the best way of dealing with the issue(s)?

In attempting to answer these questions we are forced to consider whether other approaches would be more suitable, such as writing a letter or making a number of telephone calls. If a meeting is still considered to be necessary, the issues that are to be addressed should be clearly stated and only the individuals best suited to dealing with the issues should be invited to attend.

Who should attend?

There are no regulations or rules on the number of parties or organisations that should be contracted to design and manage the construction project. Equally, there are few construction publications that give real guidance on which or how many professionals should be present in construction meetings. For reasons of economy and in order to facilitate the potential for optimum results, the number of people attending should be kept to a minimum. Early communication scholars such as Thelen (1949) suggest that the group should be just large enough to include individuals with all the relevant skills and knowledge to solve the problem. This is 'the principle of least group size' (Hare 1976). However, when groups are below five participants they may be too small to be effective; likewise when group sizes become too large members become dissatisfied with discussions. The optimum size for a problem-solving group is considered to be five (Slater 1958) and for a discussion group it is six participants (Bales 1958, Hackman and Vidmar 1970). While this may be achievable in the majority of meetings it is the site progress meeting that invariably has a much higher number of attendees and it is these meetings that can become rather cumbersome, and hence ineffective. Careful consideration should be given to the number of parties who are invited to meetings, phasing attendance to keep the meeting dynamic. Slater found that groups of five were more effective in:

- Dealing with an intellectual task involving the collection and exchange of information about a situation
- The co-ordination, analysis and evaluation of this information
- Making a group decision regarding the appropriate administrative action to be taken.

Phased participation in meetings

Professionals enter the process at different times and contribute to different extents depending on their particular expertise. It is inevitable that different professionals will attend different meetings during the course of a project. For example, subcontractors may be asked to attend meetings while their particular area of expertise is most needed (and relevant), and as their involvement in construction tasks reduces, their need to attend meetings will also reduce. When the membership varies between meetings it is likely that there will be some problems with group development. As group members become more familiar with each other they develop a better understanding of the group's interaction behaviour and regulatory structure. Members learn which members are more likely to support their ideas and which members will contest; they know whom to trust and of whom to be wary. Knowledge of other members' potential reactions enables members to use a broader range of communication techniques, being both more relaxed and more assertive where necessary. When new members enter meetings there will be a period of unfamiliarity and members will be unsure how the new member will react to different communication styles. The result of this is that when new members enter team meetings the participants are more likely to focus on task-based and factual issues, subconsciously and consciously exploring the participant's reactions before engaging in more assertive or emotional interaction.

Before sending out the invitation and the agenda for a meeting, the following should be addressed:

- Ensure that only those necessary are invited to the meeting (thus keeping numbers to a manageable level).
- Identify what resources, skills and knowledge a person will bring to a group.
- Ensure that each party's relationship with others and the tasks being discussed are conducive to problem-solving.
- Check whether an individual's status or position will interfere with the process.
- Avoid allowing political factions, hidden agendas or personal vendettas to influence group interaction or member participation.
- Where the meeting requires attendance from personnel in different organisations, attempt to control the number of representatives that attend from each organisation (thus helping to reduce the 'ganging-up' effect).

- In situations where a large number of people must attend, consider phasing the meeting so that people can leave or join when relevant (this helps to reduce wasted time and stops people contributing to a debate in which they have no stake). While phasing participation, attempt to keep groups as stable and regular as possible.
- Where direct interaction is not essential, ask potential attendees if they would be happy to send a report (and be contactable by telephone should a query arise). Individuals should not be attending if they have nothing to contribute.

Once the aims and objectives have been set and the attendees decided upon, it is necessary to set a clear timescale and agenda for the meeting.

Interaction and participation in multidisciplinary meetings

Interaction during meetings is not the same as in casual conversation. Meetings normally have an organisational purpose and restricted turn-taking procedures. Interaction is often hierarchical, more senior members and the chair often have greater participation rights, although the turn-taking is rarely fixed. The structure of the meeting is most closely associated with an open information network (decentralised), although there is the possibility that communication may not flow freely. Even without the formal turn-taking procedures some people are less willing to communicate ideas than others within organisation settings because of the influence of, for example, group norms (Daly *et al.* 1997). Some members may have a greater influence on the group interaction and decision-making than others which would mean that the meeting would not necessarily have open lines of communication. Factors such as communication dominance, influence and reluctant communicators may affect the group's interaction (Figure 12.1).

In team meetings it is important that members are encouraged to participate. Research shows that members who take part in group decisions are often more committed to the final decision and feel greater satisfaction in group processes. People are more likely to comply with a decision if they are involved in its development. Although participation does not have to be evenly distributed for the

Free-flowing information. When discussions emerge, all parties should have equal opportunity to contribute.

Human behaviour means that the strength of the network varies, the amount of information transferred depends on the parties.

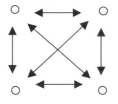

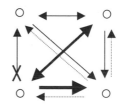

Meeting should represent a network of free-flowing information.

Some parties block others, are much more dominant, or lack confidence and are reluctant communicators.

Figure 12.1 Interaction during meetings.

group to make an informed decision, it is essential that a more dominant non-specialist does not suppress specialist contributions. When it is known that a specialist is reluctant to communicate, the chair should encourage them to take part by asking questions and prompting them for information. Reluctant communicators should be given time to speak, this may mean that others who are eager to contribute have to be controlled. The most import aspect of making complex decisions in a multi-disciplinary team is that the relevant specialist knowledge available is exploited to the full; and to achieve this requires an experienced chair.

Interaction style

The nature of interaction that occurs during meetings is a function of many inter-related variables. Some factors considered to affect the group and individual interaction during meetings are shown in Table 12.1.

Interaction styles vary from group to group and from person to person. However, researchers have found that some patterns of interaction tend to repeat themselves, and certain strategies in certain contexts appear to have greater effects on the group. Research has also shown that when several participants in a meeting share interaction strategies, one interaction style would tend to dominate to the advantage of the 'in-group' and the detriment of the others. Gorse (2002) found that the interaction during construction meetings is predominantly task-based with a small proportion of relational and emotional communication. He also found that those members considered to be more effective managers (by company directors) tend to use slightly more emotional interaction than their contemporaries. They show higher levels of agreement, disagreement, tension, tension release and also ask more questions and give greater direction to the group. Gameson (1992) found that different professionals and clients exhibit different interaction patterns depending on their particular profession and experience. Data showed that construction professionals tend to concentrate on issues most relevant to them. For example, quantity surveyors make greater use of words associated with cost and legal issues, architects concentrate on building factors, and contractors emphasise project organisation factors, such topics being more aligned with their educational and professional background. He also found that clients and professionals with the most experience tend to dominate communication during meetings.

Table 12.1   Factors affecting group and individual interaction

National and cultural issues	Personal factors
Rituals	Individual motives and goals
Religion	Personal likes and dislikes
Language	Introversion and extroversion
	Emotional awareness
Organisational and project culture	Thinking and feeling
External vs. internal emphasis	Sensing and intuition
Conservatism vs. risk-taking culture	Perceiving and judging
Reactionary vs. planning	
	Preferred interaction style
Team or group characteristics	Conscious and subconscious actions
Group goals	and reactions
Formality–informality	Task orientation
Task vs. social focus	Self-orientation
	People orientation
	Power and status within the group

Duration and time

The time when the meeting is held can have a dramatic effect on group behaviour and the final outcome of the meeting. Holding meetings at the end of a working day is not a particularly good idea because individuals are tired, more irritable and often have important social events that they wish to attend. While work may hold a higher priority than the social activity, disrupting others' schedules can cause tension. Such tension may result in confusing signals because if a person is seen to be tense, distracted or unconcerned by others it may be taken as a sign of disagreement, despite the fact that the problem lies elsewhere. Attempts must be made to ensure that all minds are fully engaged with the issue on the agenda. As 'alert indicators' we want to know when people do not agree, we do not want to be misled by frustration or anxiety shown on the face of a person who is concerned that they might be late collecting their children from school. Early morning meetings are also difficult for some people. People may be slow in the morning, tired, or have difficulty meeting the scheduled time due to traffic congestion. Careful and sensitive planning of meeting times can have a major impact on their effectiveness.

The duration of meetings is also an important factor, too frequently overlooked. When meetings go on for too long people become distracted. It is essential to build in a number of breaks during long meetings. It is also helpful to have water on the table at all times because as hydration levels in the body drop we become fatigued, tired and irritable, this is easily cured by having water available and taking breaks for refreshment.

A clear timescale

Common problems associated with meetings are that the date of the meeting is changed at the last moment and/or they invariably run over the estimated time. This is a clear indication of a poorly planned and poorly chaired meeting. We have all experienced such irritations (and no doubt will continue to do so), when in the vast majority of cases better managerial control would have prevented such situations from arising.

When meetings are changed with very little prior notice it is important to question the motives for the change. In the majority of cases the change is made to suit the meeting organiser, not those due to attend. The result is that (say) six people are inconvenienced to suit one. It is unacceptable. Meetings must be scheduled well in advance and the date and time adhered to for the benefit of all concerned. Specifying a minimum time that should be allowed for the meeting is also good practice. If an individual cannot make the scheduled meeting then another member from their organisation must be able to substitute for them and be prepared to make a contribution.

At this point we need to mention that clients are particularly prone to trying to change meetings. Some clients have little consideration for their project team and often ask for meetings to be rearranged for their sole convenience. This must be resisted and the meeting schedule retained. If the client wants another meeting then the additional cost of this should be forwarded for their consideration (once they realise the additional cost involved they soon change their mind).

Careful organisation of a meeting is critical to its success. The following points should be adopted to ensure meetings are run to schedule:

- Do not organise meetings that do not have a clear start and finishing time
- Consider adding guide times to the agenda (and stick to them)
- Keep the planned duration of the meeting to a minimum

- Consider phased meetings to allow attendees to leave and join the meeting at pre-planned times.

A typical agenda

Over time, organisations tend to develop their own 'standard' agendas for different types of meetings and/or use typical agendas published in guidance documentation produced by their professional organisations. Typical and standard agendas are very useful and provide a convenient starting point, but there are a number of dangers associated with applying them without due consideration. The first is one of relevance. Are all of the items to be discussed relevant to this particular meeting, i.e. are some of the standard headings redundant? (Invariably, the answer will be, 'yes'). The second is one of over-familiarity. Standard agendas can lead to complacency and the tendency to rush through certain parts of the agenda and/or forgetting to add an important item to the agenda. Each and every item on the agenda must be there for a particular purpose.

Agendas typically include the date, location, name of meeting, topics to be covered, the order that they will be addressed and the designation of responsibility for action during the meeting. Circulating an agenda prior to the meeting allows participants to adequately prepare for the topics to be discussed.

An appropriate location

Selecting an appropriate location is another consideration. The location may be physical, in a designer's office or in the contractor's site accommodation, or it may be in cyberspace, with everyone contributing from the comfort of their own work space, or via a video conference suite. Holding meetings on site may be helpful when discussing a building, the mere presence of the building site, drawings and site activities will help people contextualise discussions. Ideally this should be conducted in the site accommodation, however there are times when meetings are conducted adjacent to ongoing work, i.e. in the open air – acceptable in fair weather, but not conducive to a decent exchange of views when it is wet and windy. Likewise, it may be important to hold a meeting away from the bustle and distractions of the site. Informal meetings may be conducted over lunch or during some social activity such as a round of golf. While such activities may seem frivolous and expensive, if they build relationships, resolve problems and prevent difficult situations leading to a legal dispute they may be relatively cost effective in the long run. Meetings conducted via the Internet or intranet have the advantage that those attending do not have to travel to a physical location, such as the designer's office or the building site. Therefore there is a cost saving associated with the actual cost of getting to meetings and the time involved in travelling to and from them. This has to be offset against the cost of the cyberspace meeting, for example the time online, the cost of the equipment, etc., but generally there is a major time and cost saving here.

Seating arrangements and table layouts

There are many theories on the most effective layout for meetings, although in reality most are restricted by the physical dimensions of the room in which the meetings are conducted. Many of the meetings that occur during the construction phase take place in site-based accommodation, the majority of which are long and narrow. The shape of the office unit usually determines the seating arrangement of

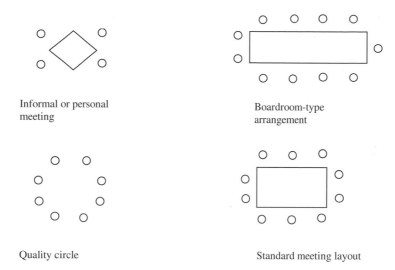

Informal or personal meeting

Boardroom-type arrangement

Quality circle

Standard meeting layout

Figure 12.2 Seating arrangements and layouts.

the meeting. Thus meetings are normally conducted around oblong tables in the boardroom or general work-type arrangement (Figure 12.2).

Some scholars believe that the presence of a table can act as a barrier to communication. It is common for groups that want more open exchanges of information to remove the table and use the quality circle arrangement. However, in business meetings the table does have a function, it is used to write on and provides a platform on which drawings and other important documentation can be placed for discussion. Boardroom-type tables often have a hierarchical function as well as a practical one with the most senior member sitting at the head of the table with other senior members in close proximity. Positions and seating configurations around a table may communicate information on position, rank, seniority, authority and power although research into site progress meetings found no obvious configurations of hierarchy (Gorse 2002). Most meetings were conducted in site-based accommodation around an oblong table; however, in a few meetings the table arrangements were somewhat irregular. On two projects the position of filing cabinets or shelves prevented a true oblong meeting table being formed and in one meeting, those participating at the meeting sat at three different and separate tables (Figure 12.3). This unusual layout meant that not all of those attending the meeting could see all of the other members, thus instant feedback through facial expression, body language and other non-verbal communication stimuli were lost. In both situations, prior attention to furniture layouts could have prevented the inconvenient layout. In terms of hierarchy, the only observation about these unusual layouts was that those most senior in the project team tended to sit in a position where they could see every member. When conducting meetings every possible attempt must be made so that all members are visible and problematic layouts must be avoided (see Figure 12.3).

The cost of meetings

Meetings can be a very effective and efficient use of resources; however, they are expensive in terms of individuals' time commitment. Given the high charge-out rate for individual's time in some instances the cost of the meeting may be higher than

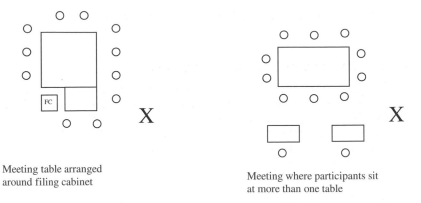

Meeting table arranged
around filing cabinet

Meeting where participants sit
at more than one table

Figure 12.3   Unusual seating layouts can obscure views and prevent non-verbal communication.

the cost of putting right the problem being discussed, i.e. it would have been cheaper to sort out the problem when it arose. If projects and organisations are to work effectively, those in control must pay real consideration to how it uses and deploys its resources. When meetings are not needed or inadequately prepared for, valuable time is wasted. However, holding important meetings and preparing for them can make them dynamic – infusing the group and motivating members into action that results in a more effective use of the resources. Again, the argument here is for careful planning and execution.

Controlling the meeting

In addition to monitoring the pace of the meeting so that it finishes on time, the chairperson has the unenviable task of trying to control the meeting, i.e. trying to enable effective communication takes place within the pre-set agenda and allocated timescale. As intimated earlier, control is fundamental. It is important to recognise suggestions, focus attention, respond to and manage emotion so that the meeting can progress in a positive manner. Boyd and Pierce (2001) asked project managers what they did in meetings and they reported that they watched individuals' body language and reacted accordingly with the aim of getting the most out of the participants, noting who responded best to gentle persuasion and who needed more aggressive action. Whether or not we need to use aggression is a subject of debate; however, emotional interaction is a natural part of group interaction. Other research has also found that professionals considered more effective use positive and negative emotion during meetings (Gorse 2002). How individuals use communicative behaviour to influence and react to others during meetings is important. When dealing with complex technical problems that involve multiple parties, interaction must develop the task and social requirements of the solution.

Interrupters

A common irritation to many that attend meetings is the interrupter. Those who are constantly interrupted may feel that they never get the chance to make a full contribution. The chair should intervene to stop the persistent interruptions (by reminding them that they have had their opportunity to contribute and it would be wise to allow others the same opportunity) and allow people to make their point.

This allows for a fair exchange of views and helps to prevent people becoming frustrated.

Essential characteristics of the chairperson

A good chairperson avoids competing with others, encourages everyone to contribute, controls aggressive and defensive behaviour and sums up clearly, stating the agreements, decisions reached and ensures meetings have the right sort of atmosphere. A good chair will also set an appropriate pace for the meeting and allow individuals the opportunity to contribute to, but not dominate, discussions.

Dysfunctional meetings

Some meetings fail to meet their aims and objectives. Valuable time is wasted and in situations where the meeting has led to ill feeling or frustration, valuable communication networks can be damaged, i.e. meetings do not always have a positive outcome. There are a number of root causes that need to be identified and eliminated. Members of group meetings often:

- Ignore group knowledge. People rarely build on another person's ideas, they wait for a chance to present their point, ignoring previous discussions
- Ignore others who don't contribute. One of the main purposes of gathering people together is to access everyone's ideas
- Concentrate too much on one chain of thought, failing to consider other perspectives. Too much attention to one item on the agenda will mean that other items are given scant attention
- Insist on discussing irrelevant issues
- Spend too little time discussing tasks, focusing on friendly interaction and relationship-building exercises
- Discuss task issues without paying attention to relationships. It is difficult to balance task and socio-emotional interaction; however, when task-based discussions become tense, attention must be given to relational interaction.

Many of the problems that emerge during meetings can be attributable to the chairperson, i.e. the manner in which the meeting is managed.

Avoidance of rituals

By rituals we mean the time-wasting (and frustrating) habit of reading through the previous minutes, going over old ground, etc. It is unnecessary. This also applies to action points that have clearly been dealt with in the reports.

Avoidance of casualness

If meetings are not sufficiently formal they invariably lack direction and hence serve little purpose. Aims and objectives need to be considered throughout the meeting, and discussions redirected if necessary. To avoid casualness it is useful to:

- Circulate the agenda and supporting reports before the meeting (thus allowing time to read, but not too much time so as to forget)
- Start on time and stick to the agreed timeframe
- Stick to the order of the items on the agenda

- Stop people talking if they have strayed from the point and/or have lost sight of the meeting's objectives
- Encourage all attendees (especially those less keen to voice an opinion) to participate in discussions equally
- Discourage individuals from dominating the discussion
- Record with due brevity the thrust of the discussion, the points agreed and appropriate action planned
- Circulate the minutes as soon as possible, no later than four working days after the meeting.

Recording and communicating decisions

By their very nature, the minutes of a meeting are not a record of everything discussed, merely a concise record of the main points, the decisions agreed and the action required. When making and recording decisions it is important that the decision is clearly understood and relevant parties are in agreement. Sometimes issues are discussed quickly and decisions made without proper consultation and agreement, which can mean that the meeting minutes are contested at a future date. It is good practice to repeat decisions and check that everyone is in agreement before making a record.

At the start of subsequent meetings members are usually asked if they agree that the minutes provide a true and accurate record of the previous meeting. When members contest the content of the minutes, the discussions are often embarrassing and tense. Starting a meeting that focuses on errors or disputes is not a particularly good way forward.

The use of IT to assist meetings and decision-making

The use of information technology to capture discussions can be helpful. With the widespread use of computers there is no reason why the meeting minutes cannot be immediately typed, printed out and signed at the end of the meeting. If issues recorded are contested they can be resolved quickly. Some meetings are conducted over the Internet or via the World Wide Web. Obviously with the use of such technology a record of interaction does exist.

As well as the Internet, other data capture techniques also exist. Software pages have been developed to help the decision-making process. Meeting-support technology, known as group support systems, group decision support systems and electronic meeting systems, is being adopted by other business sectors to increase the effectiveness in the group decision-making process. Most of these packages have been developed to encourage greater collaboration during decision-making. The meetings are facilitated to encourage idea generation, idea evaluation and prioritisation or voting on suggested solutions. Using the packages, members have an equal opportunity to contribute, contributions are anonymous (so that there is no fear of challenging others, regardless of who they are) and each person's decision is considered. The use of such technologies relies on each person in the meeting having access to a computer terminal or small keypad. Each system has a different way of operating. Some simply use the technology to anonymously show those in favour or against a proposal (using computer screens to display the results), while others allow anonymous criticism and comments to be displayed. The obvious advantages are that people are less likely to be affected by politics, power or other social influences. However, researchers at this stage are unsure to what extent the benefits gained from anonymous computer-facilitated interaction via computer consoles outweighs the benefits of a 'real' group system or social interaction that has naturally evolved (Scott 1999).

Decisions made outside meetings

Many decisions are made outside the meeting forum, either before people start the meeting, or in discussions after the meeting. These tend to be face-to-face discussions between two or three individuals anxious to reach consensus over a particular issue in order to present a united view and thus help avoid conflict at a later date. The importance of such interaction has led people to build in a small period of social time before and after meetings. When meetings become long, small breaks can also provide short informal periods where one-to-one discussions can take place and sensitive issues resolved (again helping to overcome conflict).

Pre-meeting discussions

Agreements are often made prior to the commencement of a meeting. Some issues will be quickly decided between the relevant parties. Often prior correspondence has been exchanged and the meeting has provided the first opportunity for the parties to meet face-to-face. Meeting face-to-face is often helpful in bringing matters to a close, important issues can be quickly resolved. Such discussions are best held outside the formal setting and the result of the discussion communicated quickly and efficiently within the meeting.

Post-meeting discussions

It constantly surprises us that so many decisions are made after the meeting has finished, usually in discussions between no more than two or three participants. Our own analysis of this tends to suggest that individuals are keen not to lose face, or be seen to stand down on a point in a meeting, but will concede the point afterwards when fewer people are likely to notice. Furthermore, people may defend their position when issues are first raised, but may reconsider the issue following the initial debate. Parties may muse over possible options, while other issues are discussed, and following the meeting may decide to attempt to resolve matters rather than allowing them to drag on and become acrimonious.

Different types of meeting

There are a number of different types of meeting that routinely take place during the design and construction process. Some are relatively informal and are organised on an ad hoc basis; others are formal and scheduled in accordance with programme and/or contractual demands. These can be divided by the function of the meeting into the following types:

(1) Client briefing meeting(s)
(2) Design review (formally scheduled)
(3) Design team meetings, i.e. with other consultants
(4) Pre-contract meeting (formally scheduled)
(5) Site progress meetings or project team meetings (formally scheduled) – management and design team meetings
(6) Constructor team meetings, i.e. with sub-contractors

(7) Hand-over meeting (formally scheduled)
(8) Feedback meeting(s).

These are discussed in more detail below.

Meetings with clients

Meetings with clients conveniently fall into two categories, those concerned with client briefing and those concerned with matters of progress. They serve very different functions and should be structured and managed accordingly. Briefing is a process during which the client's requirements are explored, questioned, refined and eventually recorded in the form of a written brief. It is a crucial phase in the design of a building and the atmosphere should be open and supportive thus allowing a free and frank exchange of information. These meetings should be structured enough to allow all the pertinent issues to be explored, but informal enough to allow the pursuit of alternative views and options. The main aim here is for the brief-taker to achieve empathy with the client and therefore be responsive to the messages received. Progress meetings will be more formal and are concerned with the designer or project manager reporting on matters of progress.

Design reviews

The formally scheduled design review is an important component of quality management systems. Reviews are scheduled to take place at pre-agreed stages in the development of the design so that a formal assessment of the design against the client's brief, organisational standards and regulatory requirements can be made. The reviews are planned events that form an important part of the project programme and the project quality plan. In addition to the client, designer and project manager, all consultants who have contributed to the development of the design should be present to discuss, agree and 'sign off' the design. Thus the design review presents a series of gates in the design process through which the project cannot pass without a thorough check from both the quality manager and the planning supervisor, as well as the approval of the client. These meetings provide an opportunity for interpersonal interaction and help to ensure that all parties are aware of the design's development and the implications of future decisions. Reviews are a tool for ensuring a full understanding of the information available to the project participants at significant points in the project's development. It is a good way of helping to detect errors and omissions while also allowing another avenue for the incorporation of expert knowledge and feedback from related projects. Design reviews have been found to work well because they bring people together to communicate, share views and agree a course of action in a supportive environment. As such they serve to strengthen relationships and build a positive project culture that embraces collaboration and ownership of the product. Reviews by their nature are relatively formal and must be structured to ensure all salient points are dealt with in the meeting. However, the ability to develop a relatively relaxed and informal atmosphere, thus allowing for a frank and direct exchange of views, is crucial to their effectiveness. Reviews should address the following:

- Design verification
- Design changes
- Statutory consents
- Health and safety
- Environmental impact

- Constructability and disassembly strategy
- Budget and life cycle costs
- Programme
- Risk
- Communication routes.

Design team meetings

In addition to the formally scheduled design reviews it will be necessary to arrange a series of design team meetings (formal/informal). These may involve individuals from the same organisation only, or involve invited representatives from other contributors to the design, i.e. other consultants, manufacturers and specialist sub-contractors. Their function is primarily to exchange information and agree a suitable course of action. Whether or not these are formally recorded in meeting minutes will depend upon the formality of the meeting and the demands of the quality management system being used.

Pre-contract meetings

Prior to commencement of work on site it will be necessary to hold a pre-contract meeting to:

- Ensure that the contractor and other members of the project team understand the project requirements
- Check contract documents, i.e. check that a contract is in place and signed
- Check design and production details for completeness of information
- Request clarification and further information by set dates
- Ensure that proper records are kept and contractual obligations followed
- Agree all costs and timeframes.

Site progress meetings

Frequency of formal site progress meetings will be discussed and agreed at the pre-contract meeting. The progress meetings are usually attended by the client (and/or the client's representative), the designer, structural engineers, and other consultants, the clerk of works, project manager, contracts manager and contractor (or contractor's representative). For practical purposes meetings should be scheduled on a regular basis and held on the same day and at the same time, be it weekly, bi-weekly or monthly. The period between meetings is usually determined by the size and complexity of the scheme, for example, a complicated refurbishment project may need more frequent meetings than a relatively simple new industrial unit. In situations where a project is being fast-tracked the meetings will need to be held more frequently to coincide with the issue of additional information. Site progress meetings are invaluable in order to control progress and resolve any problems and will be used to:

- Compare scheduled progress against actual progress and if necessary agree any action to bring the project back on target
- Discuss problems like delays or substandard work that may affect the quality, cost or timing of the project
- Ensure that sub-contractors agree any action necessary so that they can fulfil their contractual obligations

- Check that any additional work or variations are confirmed in writing and that work is agreed and recorded.

Contractor meetings

There are a number of different types of meeting that routinely take place during the construction process. These include project team meetings, sub-contractor meetings, project initiation, snagging meetings and hand-over meetings. The meetings where the management and design professionals are present during the construction phase are normally called progress meetings or management and design team meetings. The project manager for the main contractor will hold regular internal meetings to ensure that each aspect of the contractor's work is properly managed and controlled. Amongst other things, the meetings will be used to:

- *Control*   To record and review progress against schedule, check cost and earned value against targets set, check against benchmark standards and ensure quality is maintained, and update any records as necessary.
- *Co-ordinate*   To organise work packages, fix intermediate activities and start-times, clarify and detail methods of work, ensure preceding operations are complete allowing succeeding operations to commence, ensure adequate resources are deployed to allow operations to take place and ensure a safe working environment.
- *Administer*   To ensure records are maintained and updated, to identify information requirements, to agree variations, and to ensure safety and quality manuals and procedures are followed.
- *Foster team relationships and motivate*   To deal with human resource issues, establish and develop relationships, resolve problems, clarify responsibilities, resolve confrontation and foster group cohesion.

Implementing decisions is the most important phase in the meeting. Once the decision is made, the person chairing the meeting should ensure the necessary action is clearly understood and followed. Decision-making enables the construction process to move forward. Problem-solving within meetings during the construction phase falls within the technical and operational decision categories. Technical decisions will affect the final product; operational decisions will impact on the process. Thus, the nature of professional interaction during meetings has the potential to affect the quality of the final product and its delivery.

Hand-over meetings

The hand-over meeting will be formally scheduled for the end of the contract and should be an opportunity to celebrate everything good about building design and quality construction. This meeting provides an opportunity for the client to formally take possession of the constructed works and to thank everyone involved in the process. For the contract administrator and others this can prove to be a very stressful time as there is no guarantee that the client will be entirely satisfied with the finished result. Good hand-over meetings are well-managed events in which the client is guided around the building and any work still to be completed pointed out, the reasons for non-completion explained and the timeframe for sorting out the problem clearly discussed.

Feedback meetings

Feedback meetings should take place at the end of the project or at the end of a phase of a project. While such meetings are often avoided they are important for identifying problems, success stories and identifying lessons that can be learned for the next phase of the project. Feedback meetings are also useful for both gathering information and maintaining relationships. As well as indicating the value of others' input by listening to them, responding to the information by making improvements in future projects or phases strengthens working relationships. It is important to bring projects to a positive close. Feedback meetings and techniques can be applied to various situations, for example:

- Sub-contractor and contractor – debriefing sessions used to identify lessons learned, e.g. identify co-ordination problems and how they could be overcome
- Sub-contractor and contractor and architect – identify technical problems that could improve future buildability
- Learning from building users – post-occupancy evaluations can be very useful in identifying user needs and how the most effective use of the building can be achieved
- All designers and managers at the end of a project – a 'post mortem' can be conducted to identify any problems that emerged, and make suggestions for avoiding or reducing the impact of the problems in future projects.

Making meetings dynamic

From our research and investigations of management thinking there are a number of points that we believe will help make meetings much more effective. The final part of this chapter provides a few helpful directives.

Attendance

In order to ensure people attend meetings:

- Let those invited know why their contribution is important – give them a task
- Keep to the scheduled date. Fix the date as early as possible and remind people of the date
- Send a number of prompts that help remind members of the meeting and their responsibilities, e.g. meeting invitation and date, meeting agenda, highlight items on the agenda that the individual should consider. Using a combination of media can make such reminders and prompts even more effective, for example, a letter followed by email and telephone message, and, if the opportunity arises, mention important aspects of the meeting in a face-to-face conversation.

Preparing the agenda

Preparation for the meeting starts well before the meeting itself and it is important that people give thought to the issues before they arrive.

- Contact those invited to the meeting. Identify the reason for the meeting and ask them if there are any issues that they wish to be discussed
- List the issues collected under common themes
- Avoid waiting until the meeting to deal with such issues under 'any other

business'. When matters are raised under the heading of 'any other business' members are unprepared and it is unlikely that any decisions made will be properly informed. It would be more appropriate to change the section named 'any other business' to 'matters for next meeting'. Such action may encourage those attending the meeting to contact the chair prior to the meeting to ensure that items which they wish to be addressed are specifically identified on the agenda and circulated.

Making people act on decisions

It is common for meetings to identify issues and then decide on action to be taken; however, in many observations of meetings, the specific action agreed is not undertaken. To help motivate people to take responsibility and act on it:

- Discuss issues thoroughly
- Identify action and responsibility
- Confirm that the action is agreed with the person responsible
- Ask the person responsible to specify a time when the action is to be undertaken and completed
- Record the action, person responsible and the date that the action will be undertaken and completed in the minutes
- If the subsequent action is not delivered by the specified date, enquire as to why it was not delivered, and ask for a new date of delivery. Both the previous and new date should be recorded in the minutes. If parties continually fail to deliver, the meeting minutes become embarrassing for the individuals and provide a strong supporting evidence for employment, contractual or legal disputes. However, the act of recording dates agreed and any slippage often results in action and prevents issues developing into disputes.

Conducting meetings

- Start on time, do not allow those less organised to eliminate or disrupt a prompt start
- Stick to the agenda, focusing on one issue at a time
- Encourage participation – differences of opinion are useful; ask for information, opinions and suggestions, support challenges and counter arguments, ask for different views, ask questions – ensure that people have given consideration to their proposals. Don't allow people to make frivolous statements that have no real substance
- Control dominant members – thank dominant members for their contributions and ask if anyone else has a view on the topic
- Control emotions. It is good to allow people to show some emotion; however, too much laughing and joking can make people blasé, too much conflict can result in uncontrollable tension
- Ask for agreement on a topic, but don't assume that silence signals agreement, encourage those suspected to have different opinions to voice their concerns
- Bring matters to a close – give suggestions, ask for agreement and make decisions; it is important to close discussions
- Make people accountable, agree responsibility and timeframe
- Direct the group to the next issue
- Set the time for the next meeting
- Finish on time.

Dynamic developments - Phase one
Park Road
Progress meeting No. 3
11/10/2004

Present at meeting

Neil Gorse	Dynamic developments	Jeffery Hobday	CAG Constructions Ltd
James Barker	A. Smith Associates	Ruth Nicholls	TOP Architects
Lee Dickinson	A. Smith Associates	Garfield McIntosh	TOP Architects

Distribution (other than those identified above)

Annie McTaggart	ABC Property Developers
Iain McKinney	ABC Property Developers

Meeting objective: To report progress, identify and resolve management and design problems

	Responsibility for action	Date due	Date completed
1. Matters carried forward from previous meeting The minutes of the previous meeting were discussed and the following action points raised			
1.1 CAG (JH) confirmed that they had received the ironmongery schedule for the external doors.	TOP	5/10/04	5/10/04
1.2 ASA (JB) stated that they had received TOP fax regarding alternative canopy details, and had responded with a list of queries. Initial date required 6/10/04, fax received on 7/10/04. Response required to queries now required as a matter of urgency.	TOP	3/10/04	
2. Contractor's report	Responsibility for action	Date due	Date completed
2.1 CAG issued their contractor's report no.5 showing progress up to 28/9/05.			
2.2 CAG reported that they were currently 6 days behind programme; however, they were confident that they could complete by the contract completion date of 10/11/05. A revised programme will be issued.	CAG	5/10/04	
2. Structural engineer's report	Responsibility for action	Date due	Date completed
2.1 ....*Plus add in sections for other consultants*			
3. Health and safety report	Responsibility for action	Date due	Date completed
3.1			
4. Quality issues	Responsibility for action	Date due	Date completed
4.1			

Matters for next meeting (or Any other business)

Date of next meeting: It was agreed that the next meeting would take place at 2.00pm on 25th of October 2004

Signatures: _____

Figure 12.4   Typical example of meeting minutes.

Meeting minutes

Typical information that should be included on the minutes of a meeting:

- Name of the project
- Type of meeting
- Meeting number
- Date of meeting
- Identify those present at the meeting
- To ensure that matters are dealt with systematically divide the meeting up into sections. The sectional headings can be used to ensure a specialist is given an opportunity to contribute, or that specific or important matters are dealt with
- Identify responsibility for action and date action required
- Date of next meeting should be fixed
- Any other business should really be matters to consider at the next meeting. People do not have time to consider points properly when issues are raised as any other business.

(See Figure 12.4.)

13 Conflict management

Conflict is an inherent feature of work groups with an important role to play during problem-solving processes. With so many different participants involved in the design and construction of buildings it is very likely that there will be some form of conflict during the life of a project. Conflict, no matter how minor or major it appears at the time, must be addressed and resolved quickly to enable relationships to remain stable, thus allowing the project to continue. If problems are not dealt with quickly and professionally the situation can quickly spiral out of control and lead to disputes. Considerable energy, therefore, needs to be devoted to the management of conflict. Here we deal with sources of conflict and look at conflict management techniques from a communications perspective.

Conflict in work groups

One strongly held belief is that the temporary and multi-organisational nature of construction projects is responsible for the endemic levels of conflict so frequently reported. For example, spats between architects and planners or architects and contractors are often reported in the trade press. Sometimes these incidents are of a serious nature, although more often than not the conflict is merely the result of individuals holding different positions (e.g. planners resisting development proposals, architects trying to get approval). Outside the construction sector the occurrence of conflict within groups is not unusual, indeed it is an inescapable feature of social life. In commercial environments with competitive pressures and changing technical requirements conflict is relatively commonplace and general management texts have long recognised the need to manage conflict. We should not expect construction to be any different in this regard. Construction relies on the co-operation and integration of key specialist organisations and the potential for conflict exists within each and every project. Numerous reports (Simon 1944, Phillips 1950, Emmerson 1962, Higgin & Jessop 1965, Latham 1993, 1994, Egan 1998, 2002) attribute many of the problems to the separation of design and production activities. They see conflict as an unwanted characteristic of construction that adds unnecessary cost to projects, yet research has shown that certain types of conflict are beneficial to the development of projects. It is not necessarily conflict that is the problem, rather it is the indiscriminate (or poor) management of conflict that causes the difficulties and leads to lengthy disputes, which are costly to resolve. It follows that there is a need for all involved in the development process to recognise the signs of conflict and manage conflict to the benefit of the project. The ultimate aim is, of course, to prevent the conflict escalating into a legal dispute.

Conflict

Conflict first emerges when an individual feels that someone else has frustrated, or intends to frustrate some concern of theirs (Hargie *et al.* 1999). This perception of conflict can result from differences of opinion, simple misunderstandings, mistakes and/or fundamentally different requirements. One way of viewing this would be to see it as a breakdown in communications, an inability to explain or direct at the appropriate juncture, thus leading to frustration. Conflict may also develop because the various stakeholders in the process have their own agenda (which is rarely communicated to others). Many of the parties are competing for the same business, thus the competitive relationship between the professionals can be volatile and adversarial, a point noted earlier in the book. But not all conflict is detrimental. Functional conflict is a term used to describe conflict that may be beneficial and lead to the resolution of differences, and can be viewed as 'creative' conflict. Engaging in functional conflict management strategies can help to resolve technical and organisational differences and help to achieve optimal project outcomes. In contrast, dysfunctional conflict is a term used to describe conflict that leads to disputes and is anything but creative. It follows that dysfunctional conflict should be reduced and functional conflict should be managed (dare we say encouraged).

Before discussing issues specific to construction it is first necessary to be clear about the nature of conflict. Figure 13.1 identifies a number of different events that may result in the development of conflict. When conflict develops, the parties involved will engage in discussion and negotiation as they defend their beliefs, and emotional interaction occurs as a result of tension between them. If the conflict is not resolved and tension is not dispersed, business relationships may be threatened or weakened. Where parties feel that they are no longer able to work with the other party the business relationship is liable to break down completely.

Conflict can be embraced and managed, or attempts can be made to avoid its occurring in the first place: essentially different management approaches. No matter how minor the conflict, it will bring about emotional tension, which can be difficult to deal with. Not all of us like to be in confrontational situations and so prevention is usually the better option; however, this should not be confused with our tendency to ignore the issue. In an attempt to avoid emotional encounters

TRIGGER EVENT   **CONFLICT BEHAVIOUR**   **CONSEQUENCES**

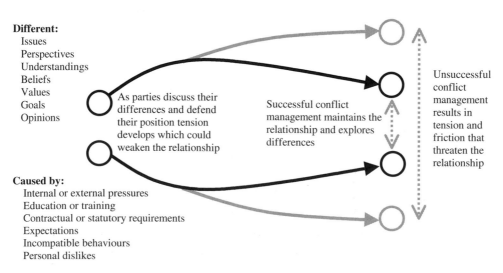

Figure 13.1   Model of conflict development.

managers may try to evade disagreements with colleagues, gloss over differences of opinion and dodge awkward situations. This 'head in the sand' approach may make them feel less threatened, but it is likely to compromise the performance of the project and lead to more serious problems further down the line. Instead, we must be prepared to listen to different perspectives and consider issues that may have remained undisclosed. If the conflict is not dealt with it can, and usually does, spiral out of control. Where issues remain unresolved the aggrieved parties rally support, seek legal advice or take some other form of action that serves to disrupt the project further.

Conflict and dispute

It is becoming common to draw a distinction between conflict and dispute. Conflict exists where there is an incompatibility of interests, in this context conflict can be managed, possibly to the point of preventing it from leading to a dispute. Disputes tend to occur as a result of conflict escalation. In situations where parties involved in a disagreement find that they are unable to resolve their differences they enter into a dispute. Disputes must be resolved and usually require intervention by others with different skills. Interestingly, once a conflict becomes a dispute and gets into the hands of third parties, individuals are judged by what they have written down (or drawn), i.e. what is recorded. It is rare for any recordings of conversations to be used as evidence. It is because of this litigious climate that we tend to communicate in a defensive manner, always wary that we must protect our position should things go wrong. It is a vicious circle that must be broken by engaging in more open communication within a more trusting and collaborative environment.

Figure 13.2. illustrates the problem of not dealing with conflict early. As those engaged in conflict move from the extreme of conflict management to the extreme of dispute resolution the involvement of third parties increases, as does the

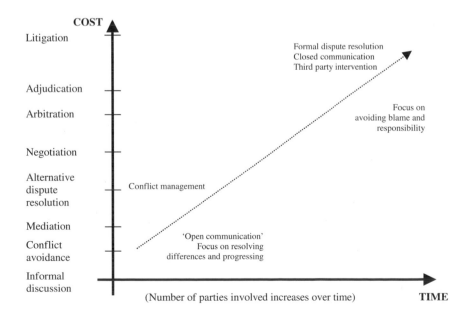

Figure 13.2 Conflict management strategies against time and cost.

potential cost of the original conflict. Projects are delayed and costs increase as the dispute becomes more serious. The involvement of third parties does little for the performance of the project.

Different types of conflict

One of the simplest classifications of conflict is provided by Burgoon *et al.* (1994). They identify five types of conflict, which are summarised below.

(1) *Real conflict* Conflict which results when goals or behaviours are incompatible. Such conflicts emerge in response to a competition for limited resources, or win–lose situations. When one party gains from the situation, the other will lose out.

(2) *Artificial conflict* While the parties may initially believe that their behaviour or goals are incompatible, both are able to fulfil their needs without the other compromising their position. The use of compromise or co-operation can help all parties to achieve their goals.

(3) *Induced conflict* An individual or group may create conflict for a specific purpose. For example, a project leader may emphasise the fact that they are in competition with others within the same organisation. The induced element of competition may help the team become more cohesive, working together against others. Conversely, it may result in members becoming defensive towards other groups, preventing effective information transfer and knowledge sharing.

(4) *Violent and non-violent conflict* The easiest distinction here is that non-violent conflict uses rhetoric while violent conflict makes use of force. While both methods can be used successfully to achieve objectives, violent conflict within organisations should be discouraged and dealt with firmly and rapidly if it occurs.

(5) *Direct (face-to-face) and mediated conflict* When conflict emerges in direct or face-to-face settings it is a result of differences of opinion between those directly engaged in the discussion. Usually, it is only after a period of discussion that a third party (mediator) is invited to help group members resolve their dispute. Third-party intervention may simply involve a mutual colleague or friend, or could be more formal, requiring the intervention of an independent mediator.

In addition to the classifications listed above, each individual within an interpersonal or group relationship will have strongly held principles and beliefs. Conflict may occur when groups operate in a way that may contravene an individual's values, for example, an individual may refuse to manage the construction of a cosmetics factory because of his or her beliefs in animal rights. For designers, conflict may be linked to their particular design values (which are being compromised), while for construction managers it may be linked to their managerial style (which is not being followed).

Advantages and disadvantages of conflict

As noted earlier, conflict can be functional (natural, constructive, creative) or dysfunctional (unnatural, destructive, non-creative). Functional conflict is described as the intended or actual consequence of the encounter resulting in stronger participants benefiting from the clash. Dysfunctional conflict is where a participant enters into the encounter intending the destruction or disablement of the other party.

Benefits gained from conflict include an increased understanding of issues and opinions, and greater cohesiveness and improved motivation (Ellis and Fisher 1994). Through argument, challenge and conflict group members are forced to see that others hold strong and defensible positions. Challenges to proposals mean that members have to defend and justify their ideas, which can help to expose key issues and areas of misunderstanding. Groups that experience tension and conflict and then work through these experiences often feel closer and stronger. Designers are familiar with this, defending their proposals by way of a design critique in the office prior to modification and later presentation to an external body, such as the client.

Disadvantages include decreased group cohesion, ill feelings and destruction of the group. If conflict goes on for too long and is not resolved, it will decrease cohesiveness within the group. Conflict between people can be distasteful and personalised, having little relevance to the task or problem, and groups that do not work through and hence resolve the conflict will fall apart. Conflict that develops into a major dispute will be resource intensive, expensive and have little to do with the initial disagreement.

The conflict life cycle

The manifestation, development and results of conflict are often considered in the following stages: disagreement, confrontation, escalation, de-escalation, conflict resolution and conflict aftermath, as illustrated in Figure 13.3.

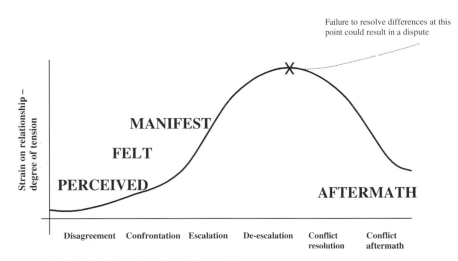

Figure 13.3   Conflict life cycle and resulting tension.

(1) *Disagreement*   As team members discuss task-related issues it is inevitable that at some point members will have a difference of opinion. Sometimes this is expressed to others, sometimes it is withheld. In multi-disciplinary teams when members are not confident to state their opinions, the potential for a solution that truly embraces all of the specialists employed may be lost. It is important that disagreements are voiced, although, it is expected that some tension will emerge from any critical debate that follows. At this stage, members of the group must decide whether there is a real disagreement or

misunderstanding. Further explanation of issues to clear up any ambiguities is essential at this point. Pondy (1967) notes that conflict at this stage can be either *perceived* or *felt*. Sometimes conflict may be perceived when no real conditions of conflict exist, or conflict may exist without the parties perceiving any conflict.

(2) *Confrontation*   If the issues being discussed are incompatible the confrontation stage begins. While attempts may be made to communicate beliefs and facts rationally, people often feel strongly about their position and consciously or subconsciously use emotional interaction to demonstrate this which results in increased levels of tension and anxiety. At this stage the conflict is said to 'manifest' itself in a variety of ways, through overt disagreement to emotional tone and expression, defensiveness, withdrawal, open aggression and violence.

(3) *Escalation*   Members become more committed to their position, tension builds and new issues emerge. At this stage the conflict (not the issue that brought it about) becomes the focus of attention and people seek to protect their self-image rather than continue with the original topic of debate. If this stage in the conflict is not effectively controlled the result will be an increase in mis-understanding, tension, distrust and anxiety. The potential of a dispute occurring is significantly increased.

(4) *De-escalation*   People will try to resolve conflicts for many reasons. For example, conflict can be frustrating, emotional debates are tiring, we may be uncomfortable with high levels of tension, others may become frustrated with the debate and people will want to bring matters to a positive close. Members directly involved in the conflict, or other group members, may try to resume rational discussions. By reiterating common goals, encouraging members to calm down, asking members to apologise, praising members for positive contributions, criticising negative outbursts and re-appraising both positions, members may be able to control emotional exchanges.

(5) *Conflict resolution*   Resolution may occur through direction, where the more powerful member imposes his/her ideas on the other members. A party may withdraw from the process, or, through discussion, a member may actually agree that the other has a stronger proposal. Members may co-operate, all parties agreeing to concede points and compromise their original position. Conflict may also be removed if the group members no longer wish to work together.

(6) *Conflict aftermath*   Conflict has long-term and short-term repercussions. The short-term consequences will manifest themselves directly from the task-based decisions made and the impact of the encounter on the working relationships. When conflict is resolved in a positive manner the basis for a more co-opera-tive relationship is formed. However, if a party is aggrieved by the outcome of the conflict, and the disagreement has merely been suppressed, the remaining frustration and tension may result in a recurrence of conflict or a personal commitment for revenge at a later date. Conflict can redefine relationships and can result in a stronger more cohesive group that is better able to deal with disagreements. Conversely, they can result in disparate groups with weak relationships. Following each conflict encounter the group member will leave the situation with ideas and rules for how to deal with conflict in the future.

Managing conflict

Because of the different languages used, entrenched habits and professional rival-ries, it is inevitable that conflict will occur (no matter how good the management) in construction and therefore it must be managed. This is usually seen as a problem, but more often than not it is a benefit because it can assist in the resolution of a problem (and can also fuel creativity). Individual goals associated with performing tasks and maintaining relationships can sometimes be in conflict with other group members' goals. When this happens, groups and individuals must confront these issues to balance group and personal goals, attempting to pull the group back to equilibrium. Functional groups need to maintain relational interaction, which involves engaging in positive and negative emotional encounters, as well as task-based communication, if they are to perform effectively. It is about taking control, resolving issues and moving forward.

Conflict-handling styles

There is enormous disagreement over the effects of conflict on the group's social system. Farmer and Roth's (1998) study of meetings found that conflict emerged in all situations and although the groups handled conflict differently, most of the group members accommodated the conflict. They concluded that accommodating the desires of others within the group, rather than attempting to fulfil their own, resulted in a less than satisfactory outcome of the meeting as a whole because different perspectives were not adequately discussed. In situations where individuals were assertive and co-operative, collaborating to satisfy the concerns of all parties, delving deeper into issues and exploring disagreements, there was a more positive outcome from the meeting. Rahim (1983) identified conflict-handling styles that were differentiated by a person's concern for others (co-operativeness) and concern for the self (assertiveness), shown in Figure 13.4. Managers need to be able to recognise these and respond accordingly.

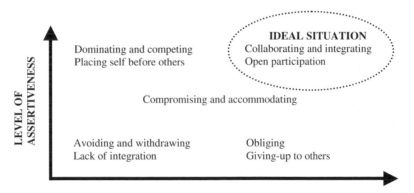

Figure 13.4   Conflict-handling styles. Source: Adapted from Thomas & Kilmann 1975, Thomas 1976, Rahim 1983.

(1) *Dominating (competing)*   This style is characterised by high assertiveness. Those who dominate have little concern for others, are selfish in attitude and have an unwillingness to consider other viewpoints. Those that adopt this perspective believe that others cannot be trusted and believe that the best approach is to be forceful. When two competing styles engage each other there is a high probability that the conflict develops into a major dispute. The dominating style results in high levels of tension, as one party wins and

another party loses, the tension between group members can remain for some time after the initial conflict.

(2) *Avoiding* This style is characterised by low assertiveness and low co-operation. Those who withdraw or suppress conflict often believe that such issues will disappear if ignored. They may believe that their opponent is too powerful or that there are few alternatives available. Such action may also be taken to reduce stress involved with the initial emotional encounter. Avoiding conflict may reduce tension at the outset, however, the long-term effects of suppression may be more problematic as the initial difficulty becomes worse through lack of attention.

(3) *Compromising* Those who compromise share. They demonstrate moderate levels of assertiveness and co-operation. Rather than attempting to find a positive solution for both parties, the emphasis is on sharing part of the burden and making sacrifices as both compromise their initial position. Although this approach can have positive effects on the group, it is not as effective as integrating-type approaches.

(4) *Obliging* One party gives up something, but receives nothing in return. The obliging party believes that there is nothing positive to gain from engaging in the conflict and that it will be uncomfortable for those involved. While this may help short-term matters there is little potential for the obliging party to gain anything from such an approach.

(5) *Integrating* The integrating style is characterised by increased participation and high levels of co-operation that results in win–win solutions. Parties explore a range of suggestions, other than just those initially proposed, searching for potential solutions that have benefits for both parties. Such behaviour may help to relieve tension and reduce the potential of escalation to uncontrollable levels. Although the level of investigation and analysis may result in relatively high levels of tension, the potential for solutions emerging that benefit all parties tends to result in member satisfaction that reduces long-term tension.

Interaction, information and conflict

Rigorous discussions between management and design professionals are essential if individual aspects of the building are to be successfully integrated and completed on time to the required standard. Interaction between the specialists is necessary in order to determine whether the individual parts will fit together and function as a whole. Most of the problems associated with construction occur during the phase when the designer and contractor have to work closely together. It is during the construction period that the abstract becomes physical and when the very different cultures of design and construction collide.

Common sense would suggest that there is a need to enter into greater discussion to avoid conflict, but research indicates that the more people engage in discussions the more the potential for disagreement. Wallace (1987) found that arguments between construction participants continue to re-emerge over a period of time, so as opportunities to interact increase so do the chances of conflict. However, Hancock and Sorrentino's (1980) study of group interaction found that where an individual had previously received support from other group members, s/he was more likely to participate in a conformist manner, thus reducing the likelihood of conflict. Many managers believe that conflict reduction is achieved through adequately informing and involving employees within the organisation; they are frequently disappointed because the increased involvement resulting from increased information can lead to greater conflict, not less.

During group development a more defined structure of interaction evolves through the group's regulatory procedures. With experience group members learn to expect conflict in certain areas and between certain members (Lieberman *et al.* 1969, Hancock and Sorrentino 1980, Wallace 1987). As the group's awareness develops the members anticipate where potential conflict could manifest itself and use supportive reinforcement interactions to overcome conflict between members. This allows participants to engage in task-related elements and control discussions with social/emotional interludes. It would seem that with the amount of information and people involved in the construction process conflict is to be expected and if managed could result in benefits such as better solutions, improved relationships and fewer legal disputes.

Interactions during the early stages of a project tend to be conducted in a relatively free and open atmosphere and the level of information available is relatively low. At the construction phase the amount of information is considerably higher and this has been found to lead to conflict. The key objective of designers and information managers should be to:

- Limit information to that which is essential
- Ensure information is complete before work starts on site
- Clearly identify any missing information and confirm a date for its completion
- Limit changes to an absolute minimum (changes mean more information, and the potential for confusion/contradiction/error is increased).

Defending resources and achieving goals

Conflict has been found to develop in multi-disciplinary design teams, as the group members discover their team objectives and then attempt to impose them on others. Also, conflict may emerge in an attempt to avoid redesigning work, i.e. as a result of resistance to unexpected changes. Inevitably situations change and evolve over time and therefore it is impossible to predict all eventualities. As situations change additional demands are placed on resources, which may simply involve a redistribution of existing resources and/or require additional resources. Both situations will involve extra work for those affected and so when changes are suggested the natural response is for parties to engage in discussions to reduce and control the impact of the change on their resources.

The management of conflict within projects needs to concern itself with the reduction and eradication of dysfunctional conflict and the management of functional conflict. Unfortunately, where there is more than one organisation each with its own goals, organisations may secure their own goals before addressing the goal of the project (Loosemore 1996). Loosemore observed that evasive and defensive behaviour often occurred during a crisis and that such behaviour attempted to reduce or minimise increased commitment of an organisation's resources to the project. This has important commercial benefits to the individual organisation, although the project gains are not so obvious.

Kolb (1992) believes that a supportive group climate should be developed so that when conflict emerges the group is able to repair emotional damage and continue with the group task. Bales (1950, 1953, 1970) believes that relational damage caused by critical discussion is repaired by positive emotional expression, showing support. Bales' (1950, 1970) research found that as groups discuss task-related issues in their attempt to resolve problems, tension between members develops. It is inevitable that conflict will emerge during this process and this needs to be managed.

Managing different perspectives through conflict

The group process of multi-disciplinary teams in relation to the interaction and compatibility of the group and individuals' goals is fundamentally different to the corresponding process in uni-disciplinary groups (Cartwright & Zander 1962). In uni-disciplinary groups the objectives of each individual are likely to be similar to those of other members, while in multi-disciplinary groups there is likely to be larger variation in objectives. The consequences of differences will result in increased difficulties in establishing a collective goal and reconciling the individual objectives with that of the project.

Historical and professional differences have led to different perceptions of social status and role definition. Pietroforte (1992, 1997) found a dislocation between the roles and rules of standard contracts and the actions of the professionals, noting that successful contracts are completed through co-operation, informal roles and rules which complement and circumvent standard contracts, possibly suggesting that professionals do not adhere to strict protocols. Wallace's (1987) research supports this view, concluding that interaction patterns of professionals are a function of group process as opposed to administrative factors imposed by contracts. Loosemore (1996) found that construction crises often resulted in conflict that discouraged collective responsibility and reduced the effectiveness of the project. The resulting conflict also generated latent tension that continued to appear cyclically throughout the construction phase. Although the crisis initially had a detrimental effect on resolving the immediate problem, paradoxically, it was found that it could also present opportunities for increased cohesion, harmony and efficiency. Individuals could demonstrate commitment and sensitivity to the needs of others, increasing cohesion and strengthening mutual trust within the project. Where cohesive teams emerged, the efficiency with which future, unexpected, crises could be dealt with increased.

Gardiner and Simmons (1992) and Gameson (1992) also found that individuals concentrated on issues more related to their professional role. For example, architects' prime concern was design quality, quantity surveyors' was cost and procurement. The different backgrounds, education and training led to different perceptions of what was of greatest importance to the project at a particular time, and this in turn led to disputes between them. When faced with a problem that requires a multi-disciplined input, two characteristics emerge. First, professionals will concentrate on the detail associated with their specialism, bringing expert knowledge into the discussion (here conflict is possible if agreement cannot be reached). Second, professionals will attempt to reduce their organisation's resource costs, i.e. limit the amount of work they have to do. This is rarely discussed explicitly, the point is that professionals will use interaction to try to influence the outcome in their and their organisation's favour. Loosemore (1994) identified two factors associated with problem-solving in construction that could lead to a defensive attitude. First, all problems involve a redistribution of resources (possibly meaning that some will benefit and some will not). Second, solutions to problems require something to change, and the act of change is not attractive to many people.

Professional misunderstanding

Misunderstanding during construction has been identified as a major contribution to legal disputes (Lavers 1992, Needham 1998). Occasionally, architects and other construction specialists may not have a full understanding of a specific construction detail, activity or service. Investigations by Lavers (1992) and Lee (1997) found that professionals may be reluctant to ask for advice, or fail to inform those relying on their services that they are unable to advise them in particular areas.

Legal disputes arising from building failure often derive from a mismatch of knowledge and expectations. Through a review of construction case law Lavers noted that it is not uncommon for a mismatch in understanding of construction or design knowledge to occur. The fact that misunderstanding occurs is not new. The Simon Report (1944) identified meetings as a forum where open communication could take place. The site meetings were seen as a place to discuss and resolve misunderstandings before deciding on an appropriate course of action. Thus conflict is expected and mechanisms need to be established to reduce anything that is detrimental to the project and individual organisations.

Relationships often degenerate and become acrimonious during the life cycle of a construction project. Hall (2001) found that open exchanges of information and sharing task responsibilities are essential for effective collaboration, while misunderstanding led to the failure of interorganisational relationships. Open exchanges do not preclude the use of challenges and disagreements to clear up misunderstandings, indeed engaging in conflict can be advantageous, encouraging greater evaluation and increasing production gains.

Barge and Keyton (1994) found that insults may be useful, although it would be unwise to advocate such a policy. They reported a situation where a member in a meeting had previously and repeatedly had his comments ignored by the chair, in a subsequent meeting he insulted and challenged the position of the chair, and in doing so became the focus of attention. The insult and personal attack (verbal) on the chair's autocratic position encouraged others to support the aggrieved member and a debate on the issue emerged. Thus, the person who had previously been refused the right to have an item discussed managed eventually to generate a discussion on that issue.

Conflict, tension and management

Bales (1970) found that as groups address problems emotions start to develop, and as a result of disagreements tension is built up between members, as they focus on the problem rather than relationships. Bales found that conflict, even when constructive, leads to tension that can damage the cohesiveness of the group and threaten group maintenance; however, too much attention to cohesion stifles constructive conflict and threatens the group's ability to solve problems. While Cline (1994) identified the importance of functional conflict to avoid groupthink and improve the decision-making process, conflict may damage relationships between group members. Conflict often emerges from perceived failure. If group members fail to meet their level of aspiration they may, or may not, try harder. Moderate levels of failure have been found to produce greater effort towards an organisation than either low or high failure levels (Hare 1976). Thus, moderate levels of conflict (being related to perceived failure) may be productive. However, negative feedback can be stressful (Shapiro & Leiderman 1967). Group members need to be aware of the development of relational and emotional tension.

Socio-emotional interaction is that which is used to form and maintain relationships. As the word implies it can be separated into two interrelated parts: the social (relational) dimensions and the emotional intensity and direction. People can attempt to form relationships by introducing themselves to others, making conversation and generally being pleasant to them. Relationships may be strengthened within groups by showing support for others, offering to help them and/or agreeing with their ideas. All of these acts are positive socio-emotional behaviours, aimed at establishing and strengthening relationships. However, group members also need to show when they are not happy with another's behaviour. Showing disagreement, antagonism and aggression are all forms of negative socio-emotional behaviour. In many ways they have the opposite effect of positive socio-emotional behaviour – they have the ability to threaten or weaken the relationship but they are equally important. When discussing issues and problems, any socio-emotional tension that develops should be removed by positive emotional acts (such as joking and praise) and negative emotional acts (such as disagreements, expression of frustration and even aggression). Bales also found that if socio-emotional issues are not addressed, the increase in tension may inhibit the group's ability to progress in its work. Groups must maintain their equilibrium, moving backwards and forwards between task- and socio-emotional-related issues.

Clearly, if groups are to achieve their goals its members must exchange task-based information, for example, identifying facts, issues, explaining the situation, presenting different perspectives, offering opinions and providing direction. To help task-based discussions members may give or request information, explanations and suggestions. However, in complex projects, as information and suggestions are exchanged differences will emerge. Group members may have different interpretations of what is right and wrong, what is of greatest importance and what factors should hold priority. As task-based issues are discussed tension between members will build and if the differences of opinion are not discussed then problems may be overlooked and the potential to reach an effective solution reduced. Using negative socio-emotional interaction during tense discussions may threaten relationships between members. Nevertheless, differences of opinion must be exposed, and the intensity of belief in the disagreement may be shown in the emotional expression. Once disagreements are exposed, group members must look to stabilise and strengthen relationships. Failure to do this may result in the disagreement becoming a major dispute, and the potential to reach a satisfactory outcome is substantially reduced. Groups that perform effectively recognise tension and disperse it with positive socio-emotional expression. Group members may recognise the differences of opinion, and identify where they agree with the other, they may compliment the members on the whole or part of a suggestion, and regardless of whether an idea was accepted, members may praise others for their contribution. Following tense debates that arrive at a solution, participants can express their satisfaction with the group, and with individual member's contributions. After particularly lengthy discussions, it is not unusual to find members making light-hearted statements or telling jokes to further ease the tension and strengthen relationships.

Bales (1950) found that some emotional acts help to develop relationships through which understanding can be conveyed; when talking, communicators acknowledge understanding by showing negative and positive emotional expression. Bormann (1996) and Trenholm and Jenson (1995) suggest that when group members respond emotionally to a dramatic situation they are openly proclaiming commitment (or not). Such expression strengthens the group's social system; group members develop an understanding of how others will react to future situations and where conflict is likely to occur. The balance between task-based and socio-emotional interaction to manage conflict during discussions in work groups is shown in Figure 13.5.

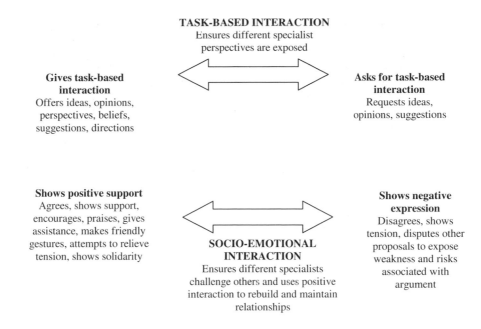

TASK-BASED INTERACTION
Ensures different specialist
perspectives are exposed

Gives task-based
interaction
Offers ideas, opinions,
perspectives, beliefs,
suggestions, directions

Asks for task-based
interaction
Requests ideas,
opinions, suggestions

Shows positive support
Agrees, shows support,
encourages, praises, gives
assistance, makes friendly
gestures, attempts to relieve
tension, shows solidarity

Shows negative
expression
Disagrees, shows
tension, disputes other
proposals to expose
weakness and risks
associated with
argument

SOCIO-EMOTIONAL
INTERACTION
Ensures different specialists
challenge others and uses positive
interaction to rebuild and maintain
relationships

Figure 13.5   Use of socio-emotional interaction to control and manage conflict and tension.

The merits of encouraging conflict

Research has found that conflict may be beneficial as well as disruptive, although few would support the deliberate encouragement of conflict. One noticeable exception is Loosemore *et al.* (2000). Drawing on work by Hughes (1994) and Gardiner and Simmons (1995) they claim that the challenge is to harness the potential good in conflict instead of merely seeking to minimise or eliminate it. Loosemore *et al.* found that the contractor's attitude and the social structure of the construction system was receptive to functional conflict, although not as strongly as they originally thought. They claim that an indiscriminate policy to remove conflict could result in a lost opportunity for increased productivity gains, which, they argue, can result from certain types of conflict. The results from this research show that changes in contractual documents and management systems could benefit the management of conflict. They also called for wider involvement in the setting, monitoring and improving of project goals. While such changes would inevitably incur a level of conflict, they would also engender a stronger sense of collective project teams.

Using Rahim's (1983) conflict measurement scale, the styles of conflict management found by Loosemore were dominated by the integrating and compromising methods, followed by avoiding and obliging, and the least used style was the dominating approach to conflict management. The dominating style is typified by an unwillingness to consider others' perspectives and develop mutually satisfactory solutions. The dominating style has little potential to generate a positive outcome. Although Loosemore *et al.* did find evidence that this existed within construction, it was the least-used method of conflict management. In contrast to the dominant method, the integrating style encourages participation and co-operation, and explores alternative solutions to a problem. This results in tension reduction and helps prevent the development of destructive conflict.

It is the dominating style which is most frequently referred to in the construction press, yet Loosemore *et al.* found that the more co-operative conflict management styles were the most common approach to resolving disputes. Clearly such approaches have advantages. However, they also found a high occurrence of compromise, and a lesser, but significant, use of obliging and avoiding conflict management styles. The compromise style emphasises sharing. Although the approach has some potential to produce a positive outcome, it usually results in a lose–lose solution where both parties give up something. The obliging style is also detrimental to the party using it, the style being characterised by one party giving up something for nothing in return. The avoiding style is characterised by withdrawal, ignorance and suppression. The long-term effects of the avoiding management style of conflict are increased levels of tension, which grow to a point where they result in dysfunctional crisis, threatening the viability of the organisations involved.

The balance of conflict styles used in the construction sector is not as bad as one might imagine. While there is a need to reduce certain types of conflict management styles, there is a definite need to encourage construction participants to be active in co-operative styles of conflict management. Such approaches do not remove conflict, they encourage open exchanges of opinion helping to identify differences as early as possible, thus allowing time to evaluate alternative solutions, and increase the potential to produce optimal solutions without a legal dispute.

Practical tools

Conflict management needs to be undertaken with sensitivity and with clear objectives. All contributors to projects have a duty to manage their own emotions to keep any conflict functional and to keep relationships harmonious throughout the time of their involvement. Those occupying the most influential positions, such as the project manager, must not only control their own behaviour but be able to recognise and respond to the signs given by others thus helping to manage potentially difficult situations. Vigilance is clearly an important skill to possess, as is the ability to communicate with people to stop them becoming frustrated and hence confrontational. Below are a number of pointers that may help with conflict management on projects, within meetings and between employees.

- Identify the outcomes required from the process
- Identify and monitor communication routes
- Monitor and check the accuracy and timing of information provision
- Respond to problems as they develop, do not ignore them
- Be assertive or passive (respond to the situation with sensitivity)
- Encourage participation, draw on different perspectives
- Stick to the main issues, avoid making issues personal
- Use emotion to show importance (however, attempt to control emotional intensity)
- Use positive emotion, showing support, agreeing, joking and laughing to ease tension and rebuild relationships
- Take breaks during meetings, allow sufficient time to cover the issues and rebuild relationships
- Following formal meetings attempt to talk to others, making sure the relationship is sufficiently maintained (or repaired)
- Ensure that all issues are resolved and that participants do not carry negative issues forward.

Whatever our approach to managing conflict it is important to recognise the importance of diplomacy. Project managers should act in a diplomatic manner and be prepared to take on a counselling role when necessary for the good of the project. Indeed, project managers have an important role to play in helping to foster an open communication culture throughout the project with the aim of resolving conflict quickly and efficiently. Encouraging the use of partnering and alliancing and hence a collaborative approach to dealing with construction projects may help in this regard.

Further reading

Kolb, D. (1992) *Hidden Conflict in Organisations*, Sage Publications, London.

14 Practical methodologies for identifying, monitoring and improving communication in practice

Throughout the text we have made reference to research and have claimed that researching communication in construction is particularly challenging. Here we look at the reasons why research into communication behaviour in practice is so important for the effective management of projects. We start with the growing need to carry out meaningful professional updating and explore some communication-related issues, before reviewing a number of practical methodologies via a case study. Some of the slightly more academic methodologies are also addressed before concluding with the issue of dissemination, i.e. the communication and application of knowledge into everyday practice.

Why communication research?

Earlier we discussed and summarised some of the published research into construction communications, concluding that more research is needed. Factors such as fragmentation, transient labour, project-specific focus and the general lack of time to reflect on the process are just some of the factors that hinder investment in research, factors that also affect the efficacy of communications. We have also emphasised the need to evaluate and re-evaluate the way in which we communicate within our organisational setting and within the temporary milieu of the construction project. To do so requires two important commodities: first is the time to carry out the necessary research (with the word 'research' used in its widest sense) and second is the determination to carry it through. It is only through research that we will get to understand complex interactions and the associated communications associated with design and construction projects. It follows that we need to look at how best to conduct and use research findings to benefit individuals, organisations and projects. We can use the published research of others, although given the specific nature of communications to particular organisations, projects and individuals it may be more enlightening to generate our own areas of enquiry. Indeed, with growing emphasis on professional updating and lifelong learning the opportunity to engage in applied research in the business environment is there to be seized.

Professional updating and evaluation

Towards the end of the twentieth century the issue of professional updating through continuing professional development (CPD) became a mandatory requirement for professionals working in construction. Implicit in being a professional is an undertaking to continually reflect and update our knowledge and skills

– a process of lifelong learning – although it is now an explicit requirement of professional institutions.

Professional institutions currently police CPD requirements by asking members to record the amount of time spent on developmental activities. A typical minimum is 35 hours per year, which roughly equates to one week per year or 45 minutes per week. Some individuals will do the minimum required to satisfy their professional institution, while others will spend considerably more time than that recommended as they seek to enhance their skills and upgrade their qualifications. Since it is the development of new and transferable skills that will help individuals to get promotion or to move to a new employer, the motivation is there. Once again, the determining factor is time. Busy professionals continue to have difficulty finding the necessary time to devote to CPD and usually find that the majority of time has to be devoted outside of their normal working day, i.e. in their leisure time. The debate as to who should bear the financial cost (time and course fees) continues to rumble on. At one extreme, the argument is that employers should pay all of the associated costs because it is their organisation that will benefit. Occupying the middle ground, the view is that a contribution from both employer (fees) and employee (time) is a fair solution. The other extreme, held exclusively by employers, is that professional updating is entirely the responsibility of individuals within their organisation. Our advice would be to pick employers with care. A mean-spirited attitude to CPD is usually indicative of a poorly managed organisation, an organisation that does not care about the development of its staff or the collective development of the organisation.

Continual reassessment and updating of skills through educational and training programmes is important for all members of an organisation, regardless of their particular position in it. Fair distribution of funds and opportunities is an ethical approach and one that also helps to maintain a balance within the whole organisation. How an individual decides to update and develop his or her skills is a matter of personal choice and will be coloured by their employer's requirements. Before embarking on a programme of self-development it is sensible to plan the short- and long-term objectives. This is essentially an evaluation of current and anticipated future needs. Evaluation needs to be considered from three perspectives, namely that of the individual, the organisation and the project. It may be useful to concentrate on one or even two of the three areas discussed below; however, unless all three areas are tackled in a synergistic manner it is unlikely that the improvements in communications will reach their full potential.

(1) *Individual needs*   As professionals we are familiar with the concept of lifelong learning and the demands it places on us. In many respects self-evaluation is the easiest method, simply because it is in our own control. Engaging in reflective practice and undertaking formal (re)training courses may enhance self-development. These may last half a day, a full day or a couple of days and may cover a wide range of communication-related issues, examples include:

- Letter-writing
- Assertiveness training
- Presentation techniques
- Managing groups/teams
- Managing meetings
- Dealing with customers, etc.
- Conflict management techniques.

Self-development may also be enhanced through research, namely:

- Undertaking a masters degree
- Research programmes (masters by research, MPhil, PhD).

(2) *Organisational needs*   Organisational evaluation can take a number of different forms. Well-managed organisations have a comprehensive staff development plan and the resources to implement it. Human capital is crucial to the running of organisations. The investment in employee training schemes which help the organisation to stay competitive can benefit both employee and employer, the philosophy being that as the individual develops his or her knowledge they will become of greater benefit to the organisation as a whole. This is the philosophy behind the 'thinking organisation'. Organisational development will rely on a combination of individual self-development (as discussed above) and formally organised staff-development sessions in which specific staff will participate. As well as helping to develop staff skills and knowledge, such group meetings are important for group cohesion and to help members recognise expertise and knowledge that exist within their organisation. Some of these sessions are delivered in-house while others require attendance at a training or educational institution. Whether these sessions are optional or obligatory will depend upon the culture of the individual organisation.

(3) *Project needs*   Evaluation of project communications is the most difficult of the three areas to tackle, but arguably one of the most important. We have already highlighted the difficulties in trying to track and monitor communication networks associated with individual projects. However, there are other, more fruitful areas that can be tackled with limited resources. For example, focusing on one specific issue (such as health and safety implementation) can help to identify areas of good and less good practice. Such applied research may benefit both the project and the participating organisations through effective feedback and incorporation of findings into current working practices.

Building alliances

Many universities and colleges are developing alliances with industry with a view to developing education and training programmes that suit the needs of individual organisations. It is well known that the majority of designers and managers are far too busy doing their job to spend time searching for appropriate information encoded in academic journals and conference proceedings. Instead reliance is placed on professional journals and books, which, by their very nature have to deal with topics that are appropriate to a wide audience. Thus guidance that may be relevant and therefore useful to a particular organisation is hard to come by. The information may be out there, but how does an organisation find it and utilise the knowledge to be gleaned with limited resources? One solution is for academe and industry to work together to share their knowledge through alliances on applied research and through work-based educational schemes such as the 'in-company' programmes. The work-based programmes can be student-centred, with the majority of the work being undertaken by the student in the workplace, or teacher-centred with a structured programme of lectures, seminars and assessment. Many successful programmes combine both approaches and aim to develop the skills of staff based around specific projects.

One of the most successful work-based learning programmes in which we are currently involved aims to utilise the skills and knowledge of a number of people from education and industry via in-company training. Employees provide education and training on internal organisation systems and management approaches. Employees are specialists in their own right, and have a lot to contribute to the general knowledge of the company, so they are helped to deliver training on internal organisational systems and management best practice. Examples include health and safety, quality, project planning, contract and financial management and knowledge transfer. Sessions are delivered by members of staff, or, where sufficient expertise does not exist within the organisation, experts from education are employed to deliver and/or contribute to particular sessions.

Through collaboration a bespoke programme is designed which incorporates a series of lectures, seminars and workshops. At the end of the programme the delegates who attend the sessions are asked to make a presentation and complete an assignment (report or essay) with the aim of helping to disseminate new knowledge. The university marks the work and provides feedback to the students, issuing awards to successful students. Some participants become so involved in their chosen subject that they become the organisation's expert. Where new expertise emerges it is used to deliver and improve the next generation of courses, thus improving the organisational knowledge base of the company. Thus the process is circular, building and enhancing the organisational knowledge base. Through knowledge transfer a learning culture is fostered that will help with the continued identification and application of knowledge.

The system has been particularly successful with many delegates gaining professional development certificates, postgraduate diplomas and masters degrees. The greatest success of the scheme is that it develops individuals and encourages a network of employees who are able to locate information and seek out experts within their company. We believe this is one of the few examples of a real 'learning organisation' where mechanisms have been formally established to develop the individual and make effective use of the knowledge within the organisation. Programmes such as this also benefit the educational establishment because knowledge of current practices can be incorporated into other educational programmes, thus helping to inform the student learning experience.

To summarise, work-based development programmes seek to:

- Identify shortfalls in knowledge
- Encourage staff to share their knowledge (e.g. through seminars)
- Identify new areas of knowledge to develop
- Better understand communication channels and manage them more effectively
- Identify opinion leaders within the organisation
- Identify change agents and work with them to assist in technology transfer
- Recognise the influence of organisational gatekeepers.

Recognising learning styles

Before embarking on a programme of self-development it is essential to recognise how we best learn, thus helping to maximise the limited time available and also helping to ensure a successful outcome. A whole raft of educational and training programmes are offered in different formats, ranging from the familiar day and evening classes offered by local colleges, universities and training companies, to open, flexible and distance learning. The choice of one over another is dependent on individual needs and aspirations. For individuals situated a long way from a university a distance learning programme may be ideal; however, just as many people prefer to do their business face-to-face, they also like the personal interaction offered by the classroom. This is particularly important when aiming to develop interpersonal communication skills and a useful first step before embarking on a programme of research. In essence, we are arguing for individuals to first have a good look at what is on offer.

Methodological and ethical issues

Construction projects benefit from communication that facilitates, rather than hinders, the delivery of the client's objective with respect to quality, time and cost. As such it is important to determine the characteristics of communication that help to bring about successful project outcomes. The effectiveness of communication in this context, as a medium to facilitate the delivery of a built artefact, should be tested in terms of the project's ability to perform within the predetermined parameters. So, rather than trying to investigate too wide an issue, which would compromise the relevance of the research, it is necessary to focus on a particular event, situation or individual.

Selecting an appropriate research method is fundamental to the development and completion of a good piece of research work, regardless of its scope. In their evaluation of different research methodologies Seymour and Hill (1993) claim that the most important question a researcher can ask is, 'what is going on here?' This is particularly true of communications research. The physical size of construction projects, the length of time from inception to completion and the intricate social networks that develop during construction projects prove a formidable, yet exciting, challenge to researchers (see case study below). It is particularly important, therefore, to clearly identify the rationale for the research and define the limits of the study. It is this criterion that will influence the research methods used to gather and subsequently analyse data.

Qualitative and quantitative research methods

The distinction between qualitative and quantitative research methods can sometimes be slightly confusing; however, from a communication perspective, if the data is transformed into numerical data, presented in graphs or statistics it is quantitative. Qualitative data does not seek to turn data into quantities, rather it helps to analyse interaction, statements and transcripts with the intention of identifying patterns, links, beliefs and trends. More specifically:

Quantitative methods of analysis

- *Statistical analysis*   When quantitative information is collected it is normally analysed with the aid of either descriptive or inferential statistics. Descriptive statistics simply segregate and aggregate the data and use various methods to present data graphically, e.g. histograms, pie charts, tables, etc. Inferential statistics use various formulae to determine the probability of something occurring, or identify the strength of the relationship between two or more variables.
- *Content analysis*   This form of analysis usually seeks to classify communication acts into categories that have common features. Once categories are classified statistical analysis can be applied to the categories.

Qualitative methods of analysis

- *Conversation analysis*   This is concerned with the contextual sensitivity of language with a focus on interaction and social action. Investigations using conversation analysis can only be pursued through intensive qualitative analysis of interaction events. Transcripts or audio recordings of interaction are required to provide the detailed data necessary for conversation analysis. The analysis attempts to understand the relationship between different events.

- *Discourse analysis*   This is a slightly broader term than conversation analysis. It involves the scrutiny of transcripts of discussions and statements. The content and the linguistic context are considered when establishing meaning and intention of the interaction.
- *Semiological analysis*   This form of qualitative analysis assumes that there is a relationship between the appearance and structure of the text and interaction and the meanings that it produces within a specific culture or context.

Ethical issues

Research, by its very nature of enquiry, is invasive and care needs to be taken by the researcher at all stages in a research project to ensure that the interests of those associated with the subject being researched are not compromised, i.e. an ethical approach is required. Given the sensitivity of research into communication in a commercial environment we must be careful to ensure that data remains confidential. Thus any publication of the data and subsequent analysis must follow certain protocols to preserve anonymity. For example, the coding up and interpretation of data must not allow others to identify features that could tie the research back to the participants and/or organisations studied. Following the ethical route it should not really be necessary to state that the work should be transparent and defensible. The reasons for conducting the research, the assumptions made that led to specific beliefs, the key issues emerging from previous work, the methods used, the problems encountered and the limitations of the work should all be clearly stated. This allows the reader to see exactly what was done and allows other researchers to repeat the work in a setting relevant to them.

Many methodologies used in the past have been unable to capture sufficient communication data to examine communication problems. Difficulties have been experienced with observing organisational communication as it occurs in real time. The first problem encountered by researchers who wish to record communication is the amount of time and energy that are often required simply to negotiate access into the project environment to observe interaction. This tends to increase as the following factors gain importance:

- The degree of invasion into a professional's environment
- The ability to accurately record communication, providing a factual record that may be used against someone in the future without compromising confidentiality
- The inconvenience of participants having to undertake additional tasks or spend additional time that may be required to take part in the research.

Obtaining permission to observe people's behaviour in detail is a major research problem. The business environments that are of greatest interest to the management researcher are often those that are most sensitive. In addition to concerns over commercial sensitivity of the data there is also the problem associated with being observed. None of us are too keen on being watched as we work, we may feel uncomfortable and naturally suspicious of the researchers' motives. Researchers have detected a certain amount of defensiveness towards them and it is clear that an element of trust needs to be built between the parties before observations can be conducted. There is an additional problem linked to ethnographic research in that we cannot discount the fact that the observer may have influenced how those being observed behave, simply by their presence.

Recording acts of communication

Communication is a complex phenomenon, the communicators simultaneously send and receive multiple signals at different levels. Signals are sent at both sub-conscious and conscious levels, yet the observations of communication are limited to aspects that are processed consciously. Interpersonal communication transmitted through expressions, sounds, actions and reactions can be observed by a third party. In contrast, investigations of intrapersonal communication, the inner thoughts and beliefs, can only be accessed by retrospective explanations, records or accounts supplied by the originator. Research may seek to classify and categorise interpersonal communication acts, leading to the development of models based on classification systems. A difficulty associated with studies of this nature is that they must transform communication that is continuous, intermingled, overlapping and rather abstract into observable phenomena. Care must be taken to select or develop an appropriate methodology that is reliable and consistent, and which can be replicated by others thus helping to determine validity.

The limited degree of conscious control that humans have over communication acts makes organisational communication particularly interesting. Speech is the product of an unconscious process. During face-to-face interaction, we cannot plan grammatical structure; there is not enough time. Speaking is one of the many things of which the cognitive unconscious takes care (LeDoux 1998). Although we cannot observe the subconscious processing of communication acts, we can observe the results: the emotions, the expressions and the speech. Yet, research on communication in organisation settings is considered by some researchers as too complex or inappropriate to model using quantitative methods. The use of quantitative methods alone offers a limited perspective because the focus of the research cannot be controlled. The unit being observed develops, changes and responds in different ways depending on how the professionals act out their roles. There is a danger, particularly with statistical methods, of becoming too focused on the intricacies of measuring, and thereby focusing attention on the classification of interaction instead of observing what is actually happening (Cassell & Symon 1994). The use of both quantitative and qualitative research methods increases the detail of the information collected thereby improving the overall methodology and hence reducing some of the research limitations.

Since there are plenty of good books that deal with the different methods of data collection we have confined ourselves to a brief overview of those used in construction communication research. The case study reported below helps to identify some of the benefits and disadvantages of a select number of methodologies.

Case study: observing construction progress meetings

Research into communication during the construction process is limited by the communication behaviours that can be observed within the parameters of the available resources. The scope of the research must be carefully designed and controlled so that meaningful and manageable data can be collected, thus research tends to be focused on one particular event or one aspect of communication. For example, doctoral research has looked at pre-contract design team communication (Wallace 1987), client briefing (Gameson 1992), interaction associated with a major sub-contract package (Pietroforte 1992), the client–quantity surveyor relationship (Bowen 1993), crisis management during construction (Loosemore 1996), the specifier–manufacturer relationship (Emmitt 1997), and the examination of group coherence in team meetings (Hugill 2001). All of these studies used qualitative approaches (diaries, interviews, observation) although Wallace and Gameson also used a quantitative system (Bales' (1950) interaction process analysis) to classify communication acts.

Diaries and interviews

Diaries are used to produce a self-report or measure of the subject's feelings or beliefs (Symon 1998). This type of methodology assumes that people can provide relatively accurate accounts of past events, and although abbreviated, they provide a source of data that is difficult to obtain using other methods (Clark 1991). Qualitative diaries were rejected by Mintzberg (1973) because they were too structured and also dismissed by Stewart (1967) after completing a large diary survey which was seen to be unreliable because the respondents were interpreting their activities in different ways. However, Symon (1998) puts forward a very convincing argument for the use of qualitative diaries as a useful and insightful tool. Loosemore (1996) used diaries to investigate communication behaviour patterns during incidents of crisis management, supported with semi-structured interviews and non-participant observation of meetings. He relied on the participant completing the dairies once a crisis had developed. The main benefit with this method was that when the crises emerged, research data could be provided without the need for researcher intervention. Unfortunately, the diaries were not fully completed by the participants, with entries declining as the workload and associated pressures on the individual increased. Although self-critical of this method of data collection, Loosemore found that the diaries did provide valid information (qualitative and quantitative) when combined with semi-structured interviews.

Observation supported by audio recordings

Audio recordings were used by Gameson (1992) to record a staged meeting (experimental laboratory research) and in the field by Hugill (2001). Hugill attended site meetings as a participant observer and, using an ethnographic methodology supported by audio recordings, attempted to study group interaction through the eyes of its members. By examining individual episodes of group exchanges Hugill was able to explain how certain utterances and interaction sequences are used to 'talk through' matters, helping the parties involved develop a better understanding of the situation. The research illustrates interesting phenomena that would have been overlooked with other methods; however, the episodic nature of the research made it difficult to generate and draw substantive conclusions. The detail of analysis attached to the content of each utterance, and the shear amount of data recorded, made it difficult to draw conclusions about the events that occur during a single meeting. Thirty hours of meeting interaction were recorded, although less than one hour was used in the final analysis. Difficulties were also experienced in gaining approval from all members present at the meeting. It is also worth noting that Hugill restricted his study to one project.

Observation using Bales' interaction process analysis (IPA)

Gameson's (1992) research was largely quantitative. Bales' (1950) IPA was utilised for classifying discussions during the first meeting of clients and professionals. Bales' IPA method offered a generic system for analysing the process of interaction rather than the content (Stone *et al.* 1966). Indeed it is claimed that there are few methods of measurement and techniques for collecting empirical readings of group interaction and behaviour that are better than Bales' IPA (Brown 2000). Previous studies using Bales' IPA have resulted in some significant findings. For example, some group members talk more than others; the people who talk the most tend to receive the most attention; larger groups tend to be dominated by one person and different people are likely to predominate in particular coding categories (Brown 2000). Gameson used Bales' (1950) IPA to determine whether certain construction professionals and clients interact more than others.

Although Bales' IPA is well used, it does have limitations, for example an inability to identify and examine the nature of the problems being discussed (Gameson 1992). Wallace (1987) also used Bales' IPA, but as part of a bespoke method. The systems adopted were not administered separately in accordance with their original protocol, but selected parts were cut, shuffled and pasted providing a unique combination of labels and categories. However, Wallace's method is not described in sufficient detail to allow it to be repeated by others.

The pilot study

As noted earlier, the use of more than one research tool to study interpersonal communication can be advantageous. Different methodologies yield different kinds of data which, when used together, allows a more comprehensive analysis of the phenomenon being studied (e.g. Fielding & Fielding 1986, Mior *et al.*, 1998). Moser and Kalton (1971) and Jobber (1991) have argued that a combination of research procedures is more useful than a single tool. It is claimed that the triangulation of data by multi-method approaches is essential in understanding organisation process (Cassell & Symon 1994). Robson (1993) suggests that different methods should be used for complementary purposes and to enhance interpretability rather than trying to 'fix the answer' as triangulation may suggest. Gameson (1992), Loosemore (1996) and Wallace (1987) all used multiple methods to enhance their studies and from analysis of the earlier research methods it was felt that the use of diaries, audio recording and observations using Bales' IPA method all had potential and were worth piloting to test their appropriateness. The pilot study was conducted over a six-month period, with the three methodologies piloted in the following order.

(1) *Diaries and semi-structured interviews*   Diary sheets were designed to collect interaction data on issues addressed during project team meetings. The data to be entered by the participant included the issues addressed in the meeting, the personnel involved and the nature of the interaction. Approval to distribute the diary sheets to five construction managers was sought, and gained, from their senior manager. The diary was designed so that it could be completed by the participants immediately after attending a progress meeting. The trial period lasted two weeks, after which time the diaries were collected and the data analysed. The intention was to interview the participants following the analysis of the data. Only one out of five professionals completed the sheets, despite the fact that all five said that the diaries were easy to use. This was disappointing because regular contact was maintained with all participants, and support for the research project had been gained from more senior professionals in the participant's organisation (who had also encouraged the participants to complete the diaries). In the case of the one participant who had completed the data sheets, the quality of the data reduced over time, questioning the consistency of the data, a difficulty also experienced by Loosemore (1998).

(2) *Observation supported by audio recordings*   Attempts were then made to gain approval to record two site-based progress meetings. First, the construction company's management team was approached (because it was they who

organised the meetings) followed by approaches to the other organisations that were party to the meetings (the contact names were provided by the construction management team). Difficulties were experienced with the amount of time required to make contact with the participants and in gaining their approval to record the meetings. For example, in one case it took over two weeks to make contact with the eight professionals due to attend the meetings. This is a difficulty with which other researchers have struggled (e.g. Hugill 1999). In the pilot study a high degree of resistance to audio recording was experienced. Two professionals refused permission to be recorded and those remaining were apprehensive because they were concerned about confidentiality. With the amount of resistance, and additional concern that recording may change communication behaviour, the methodology was abandoned.

(3) *Observation using Bales' IPA*   The final method to be piloted required an observer to attend the meetings and record (and classify) interaction using written data sheets, in accordance with Bales' IPA methodology. This method was prone to the same problem of having to contact all the participants to get their approval, which was time-consuming, but there was little resistance to an observer attending and observing meetings. Professionals were relaxed about someone attending the meetings and recording their interaction because there would be no record of what they actually said. As explained to the project participants, Bales' IPA method simply records who speaks, whom the speaker addresses, and classifies the content of what is said into one of twelve categories. The categories are distinguished by the nature of their task or socio-emotional content.

Once acceptance from all project participants had been gained it was then possible to observe a meeting. At the first meeting, the researcher was introduced to the team, the members were informed that the researcher was observing the meeting but would play no active part in the proceedings. The group members were shown a copy of the data sheet, which helped to reassure them that the nature of interaction recorded did not present a risk to them personally or to their organisation. During the first few minutes some of the professionals were inquisitive regarding the nature and use of the interaction data sheets, but once the meeting was in full progress the participants paid little attention to the researcher.

Issues for consideration

Before discussing the main research project it is necessary to reflect briefly on the difficulties experienced in trying to negotiate approval for the data collection. The sensitive nature of progress meetings and the potential conflict between those attending the meetings from different commercial organisations was a cause for concern and raised ethical issues for the researchers. Clearly, not all professionals were happy with being recorded on audiotape. Despite the researchers' integrity, they were concerned that the tapes would not remain confidential. In contrast, approval to attend (but not record) meetings was easier to gain, although still time-consuming. The use of diaries also presented difficulties. Research methods that require professionals to undertake activities that are not part of their normal duties hold a low priority and do not receive sufficient attention to be of use. From these pilot studies it appeared that the greater the perceived amount of intrusion into the social system, the greater the degree of negotiation and co-operation required by the researchers. It is clear that an element of trust needs to be built between the researcher and organisations before interactions can be observed, and a methodology adopted that does not compromise this trust. It also became clear that approval to attend meetings was extremely time-consuming and would need to be accommodated in future research proposals to ensure that the necessary resources were in place.

The main research project

Following the successful piloting of Bales' IPA the main research project could be designed and implemented. Ten projects based at sites in the north of England, employing design and build, or traditional contracts, with values ranging from £3 million to £14 million, were used as case studies. To help normalise the observations, it was decided that three consecutive meetings per project would be observed, thus reducing the potential for unusual behaviour in one meeting adversely affecting the results. This provided a total of thirty meetings. Observation took place in the contractors' site accommodation and it took two years to complete the observations. The observer sat with the participants but took no part in the discussions.

Observations were recorded using Bales' IPA technique, which identifies the communicator and the recipient of the message (discussed in more detail below). It permits classification of the statement used into either one of six 'task-related categories' or one of six 'socio-emotional categories'. Data was recorded by the observer using a prepared check-sheet with tick-boxes, which allowed for quick and efficient recording of the three variables noted above, namely; identification of the person speaking; identification of the recipient; and the interaction category that classified the statement used by the speaker. The observer made a brief qualitative note of the issue being discussed on the observation sheet during the meeting. At the end of the meeting the participants were given the opportunity to see the observation sheets and any qualitative notes made by the observer. At the first meeting the professional team did look at the sheets to confirm that no personal or confidential information was recorded. The professional team expressed no desire to see the data collection sheets at subsequent meetings.

Reflection on the methodology used

The pilot study suggested that gaining approval from all parties to observe progress meetings was likely to be difficult. For these ten projects, approval was straightforward, although very time-consuming. Once in the meetings Bales' IPA was simple to use, and more importantly those being observed appeared to be relaxed and uninhibited by the presence of the observer – a point confirmed subsequently through interviews with the participants. Consistent with Bales' IPA a reliability test was run on the data obtained and found to be within the acceptable limits set down by Bales. Running parallel with the research reported here, two postgraduate students were asked to use the same methodology to record progress meetings on different construction projects. Reliability tests were carried out and found to be within acceptable limits; furthermore, their findings were consistent with the main research, thus confirming that the methodology is robust and repeatable by other construction researchers.

The main problem with this observational tool was the amount of data recorded during the observations. This data had to be entered into a database before any analysis could be carried out and this proved to be very time-consuming and something to be considered by other researchers. In addition to the time demands, a number of difficulties arose during the observation period that may affect others using this methodology.

(1) Inconsistency of people attending meetings. From our own experience of progress meetings we expected some inconsistency in those attending. To reduce the impact and normalise such changes each project was observed three times, recording the interaction in three consecutive meetings.
(2) The construction managers, because of pressures with a particular contract, frequently changed meeting dates and times at the last minute. In the majority of cases the research team were informed of the changes, but with very little time to respond. In a couple of instances the research team were not informed of the changes, only to turn up at the agreed time to be told that the meeting had already taken place, or had been rescheduled for a future date. This posed real difficulties in observing and recording a sequence of three meetings.
(3) Change of site manager (related to point 1 above) meant re-negotiation for permission to attend the meetings, which was not readily forthcoming. Given that the site manager was new to the job and had other priorities this was not unexpected.

Despite the challenges encountered by the researchers, the work provided useful data about the interaction in site progress meetings (for full details see Gorse 2002). Given this, we have included some further information about coding systems.

Categorising and coding

The categorisation of communication is often used to allow quantities to be generated from observations of interaction; however, there are many different units of analysis that can be recorded, the most basic include:

- *Participation* Participation identifies those actively involved in the communication behaviour being observed.
- *Act* The communication act defines communication into discrete acts, e.g. verbal communication, and body and facial movement.

A core assumption of many group communication theories is that communication is the observable phenomenon binding together the systemic entities of the group (Mabry 1999). Most communication models are based on observations of external factors or indicators of communication, such as the sending and receiving of verbal and written messages, facial expressions, emotions and body language, or reactions to these messages. Observation of overt external factors of communication, identifying who makes the communication act and who the communication is specifically directed at, has been termed the 'surface meaning' of communication (Heinicke & Bales 1953). Surface meaning research recognises that there are many different levels of communication taking place, but this limits observations to those communication acts that are most obvious to the observer. The focus is on overt communication acts and their overt direction. It is the main and most obvious communication act that is recorded using this method, even though communication takes place simultaneously at different levels.

Coding

A coding system is simply a way for the researcher to view the world (Bakeman & Gottman 1997). One of the most fundamental components of the research process is discovering and documenting the discrete action-based elements – signals, gestures, units of interaction – and to specify the relationship between these elements

(Duncan & Fiske 1977). Researchers cannot observe abstract concepts; therefore, aspects of interaction must be translated into observable phenomena, using operational definitions for each of the conceptual variables (Clark 1991). It is important to establish low-level constructs, simple definitions of observable phenomena, that can be explicitly tied to the data, before communications at a more abstract, and possibly more complicated, level can be developed.

Duncan and Fiske (1977) state that observations of face-to-face interaction should be generated from a disciplined observation of categories, which require a narrow focus of attention, minimum levels of inference, discrete decisions, and moment-to-moment judgement. They also add that coders should avoid attribution of meaning and intent of the interaction. However, Bales' (1950) earlier work suggested that the researchers need to use, and cannot avoid using, their own intuition, considering statements made before and after each act. Coding systems must be objective and attempt to remove guessing what the communication aims to achieve from the data through the use of 'blind' coding.

Observers must not be biased and coding systems should not attempt to add something that is not there; however, it is difficult to code words without considering their true intended meaning. Codification is not based on words alone, but on our understanding of words when used in a context of interaction sequences supported by the emotion of the group and its individual members. So it is important to recognise discrete interaction variables within the context of the larger interaction sequence. The distinction needs to be made between systems that attempt to capture the general intent of interaction, and those that may attempt to forecast the effect of interaction.

Bales' IPA coding system

The findings of a coding system are tied to the method used to capture the data. There are limitations involved with the results, and difficulties when attempting to compare results that have been obtained from different systems. The coding system used must be clear and easy to use in the field, reliable and capable of replication by others. Several methods have been developed; however, the method that has received most attention in the social sciences is Bales' (1950) interaction process analysis (see Table 14.1). This is one of the most widely used techniques to study overt group interaction and has proved to be a consistent and reliable tool in social sciences research, but has not been used in construction, notable exceptions being doctoral research by Gameson (1992) and Gorse (2002).

IPA is used to classify direct face-to-face interaction as it occurs, summarising the resulting data so that it yields useful information. The tool provides a method for classifying interaction into one of twelve categories. These categories are divided into two groups, either socio-emotional or task-based interaction (discussed below). The technique can be used with, or without, the use of audio recording devices. This is important in construction because we have found that the majority of professionals are reluctant to be recorded but are quite relaxed about being observed. So a tried and tested method, that allows observation and classification of communication acts without the need for audio recording, can prove to be an essential tool for researchers wishing to enter sensitive environments where confidentiality is paramount. Bales' initial argument was to assume that any group is ultimately directed towards the achievement of a task. Thus, the group has a function and moves through different stages of interaction behaviour to achieve that function. The method has led to the discovery of certain interaction tendencies in different groups undertaking different activities at different times.

Table 14.1   Bales' 12 interaction categories

Id. number	Category description		
1	SHOWS SOLIDARITY Raises others' status, gives help, reward	F	*Socio-emotional acts* Positive emotional
2	SHOWS TENSION RELEASE Jokes, laughs, shows satisfaction	E	reactions
3	AGREES Shows passive acceptance, understands, concurs, complies	D	
4	GIVES SUGGESTION Direction, implying, autonomy for others	C	*Task acts* Attempted answers
5	GIVES OPINION Evaluation, analysis, expresses feelings and wishes	B	
6	GIVES ORIENTATION Information, repeats, clarifies, confirms	A	
7	ASKS FOR ORIENTATION Information, repetition, confirmation	A	*Task acts* Questions
8	ASKS FOR OPINION Evaluation, analysis, expression of feeling	B	
9	ASKS FOR SUGGESTION Direction, possible ways of action	C	
10	DISAGREES Shows passive rejection, formality, withholds help	D	*Socio-emotional acts* Negative emotional
11	SHOWS TENSION Asks for help, withdraws from field	E	reactions
12	SHOWS ANTAGONISM Deflates others' status, defends or asserts	F	

Notes
A   Problems of orientation   D   Problems of decision
B   Problems of evaluation   E   Problems of tension management
C   Problems of control   F   Problems of integration

Source: Adapted from Bales 1950

The main advantage of the Bales' IPA is its ability to examine different types of groups; it offers a generic method that is not specific to one context. One problem that exists within the system is that it classifies communication acts into either socio-emotional or task-based categories. Most statements made within work groups are in some way related to tasks and are often supported by emotional expression at the same time. When classifying interaction, systems would benefit from a measurement scale that identifies the nature of the task-based interaction and the emotional intensity with which the message was delivered. Another limitation of coding systems is that no matter how general a functional coding system may seem, and no matter how neutral the categories are, the system always represents a particular perspective and thus may not be useful in many cases. Each of the categories is developed from a vast body of explanation that is supported by rules and these are applied regardless of terminology that is specific to context. The system ignores terminology and concentrates on the intended nature of interaction. An additional system would be required if observations were to identify professional terminology or aspects that are specific to a context.

IPA has been developed further into Bales and Cohen's (1979) SYMLOG (System for the Multiple Level Observation of Groups). This is one of the widest reported systems of multiple-level observation methods, but its use may present difficulties because it requires the participants to complete a number of forms. With the increased number of people coding interactions there are increased difficulties experienced with intercoder reliability. The time for each individual member to understand and complete the SYMLOG self-study sheets is about three to four hours for a group of five, and longer still for larger groups. Despite recorded benefits over Bales' IPA, in live business environments there are real difficulties associated with applying this type of methodology. Professionals' time is very limited and the participation of every member of every meeting would be required, which given our previous experience is highly unlikely. However, during role-play exercises, or training events, such tools could be very helpful. Understanding how others behave and feel during interaction is extremely important. Using team activities to expose group members to the feelings and beliefs experienced may help develop a group that has a greater appreciation of others.

Recognising weaknesses of research

It is important to remember that research findings are merely an indication of a particular set of circumstances at a particular point in time. We cannot prove anything beyond reasonable doubt. Research results do, however, provide both researchers and the intended audience with an insight into a particular issue that was not previously available. This may serve to reinforce a previous hunch or may well bring some new knowledge to light. Whatever method or methods are adopted, as long as the protocols of the research discipline are followed and the results are reported honestly, the research may make a small, but collectively important contribution to the field, from which others can learn. A couple of issues need to be addressed here, namely the issue of the researcher affecting the behaviour of those being studied and the associated issue of the representativeness of the work.

Affecting behaviour

When observing interaction in the workplace we must remember that the presence of the researcher may affect the behaviour of those being observed. Individuals may change their behaviour (perhaps unconsciously), helping them to provide what they consider to be a suitable impression, i.e. they tend to act as they think they are expected to, rather than how they might otherwise do if not being observed. Although it may be possible to observe the subjects without their knowledge, e.g. by using hidden recording devices or being present under the guise of a normal participant, such approaches raise a number of ethical issues that may be difficult to overcome. There are many studies that have argued whether or not observation or recording interaction affects behaviour. Our own research indicates that the influence is minor; however, it can never be ruled out.

Representativeness

Problems of representativeness have been recognised where research is based on a small number of case studies or observations; however, there is little guidance of how to determine the appropriate number of cases. The length of the observation period, number of observations or number of participants involved varies depending on the nature of the study and the research method used. Research based on the analysis of four case studies is generally accepted as a minimum number for drawing meaningful conclusions. However, there are occasions when research based on one case study is useful in helping to highlight particular issues. For example, Wallace (1987) used one longitudinal study of the design process, lasting 18 months, to develop an understanding of design team interaction, and 16 cross-sectional (single) observations on different projects. Gorse's (2002) investigation of construction meetings was based on the performance of the contractor's representatives involved in 10 different projects and observed 30 meetings. All of the data was statistically analysed using a quantitative content analysis method. Hugill (2001), who undertook a similar study, observed a series of meetings recording 30 hours of discussion; however, the qualitative conversation analysis method used to interpret the data meant that only 1 hour of interaction data was analysed. With a small sample size, it is not be possible to determine whether the findings are representative of a wider population. Instead, we have an illustration of issues affecting a particular event at a particular point in time from which we can draw conclusions.

Whether we like it or not, research is limited by resources, time and cost limitations, and so researchers must be realistic when designing their projects. Occasionally it may be necessary to modify the initial objectives rather than attempting to collect huge quantities of data that may be difficult to handle and analyse. A method is required that can provide consistent data that is representative of the situation from which the data was taken.

Dissemination: a constant challenge

There is not a great deal of point in undertaking original research and then not disseminating it to a wider audience. In industry the paranoia over competitive advantage makes the chances of dissemination outside the organisation unlikely. Organisations undertake research (into markets and competitors' products) for their own use and are keen to keep those findings confidential because they are commercially sensitive. Dissemination is restricted to certain members of the organisation and may even be kept from other divisions who are seen to be competing. In contrast, academics are duty bound to disseminate their findings to a wide audience, through books, peer-reviewed papers, articles in professional journals, presentations at conferences and to industry, and equally important via incorporation into teaching material for undergraduate and postgraduate work. All of us who are involved in construction have a duty to improve the manner in which we communicate and improve our collective knowledge about this critical area. From the perspective of the organisation it is important to consider the following:

(1) Dissemination within the organisation
    In a highly pressured environment the temptation is to distribute a research report by email and hope everyone reads it. This must be avoided. Employees rarely read reports unless they have to (they simply do not have time) and even if they do there is always the danger of misinterpretation and misunderstanding. Furthermore, how does one get meaningful feedback. Instead, time must be made to present and discuss the findings of the research in meetings where individuals are invited to give their reaction and comments.
(2) Dissemination outside the organisation
    Edited highlights of research may be disseminated to other, carefully selected, organisations and client bodies. For example, other members of the supply chain would benefit from research findings into issues surrounding the effectiveness of communication routes within the chain.

Final words

Following our plea for dissemination of research findings we come to our final words of advice. If we are to better understand the nature of communications in construction it is clear to us that much more research is required and is made available in the public domain. This is an important factor in helping the construction sector to move forward and to respond to a changing environment.

We do offer one final piece of advice, in the spirit of Sir Ernest Gowers. It is worth remembering at all times that the purpose of contract documentation (drawings, specifications, schedules, models) and the interpersonal communication that accompanies any project is to get an idea from one mind into another as accurately and efficiently as possible. Easier to state than to do, but we believe it is worth making the effort.

Further reading

Gill, J. & Johnson, P. (1997) *Research Methods for Managers*, 2nd edn., Paul Chapman Publishing, London.

References

Anderson, C.M., Riddle, B.L., Martin, M.M. (1999) Socialization process in groups. In: L.R. Frey (ed.) *The Handbook of Group Communication Theory and Research*. Sage Publications, London, pp. 139–163.

Arnold, J., Cooper, C.L., Robertson, I.T. (1996) *Work Psychology: Understanding Human Behaviour in the Workplace*. Pitman Publishing, London.

Averill, J.R. (1993) Illusions of anger. In: R.B. Felson & J.T. Tedeschi (eds) *Aggression and Violence: Social Interaction Perspectives*. American Psychological Association, Washington, pp. 171–192.

Bakeman, R. & Gottman, J.M. (1997) *Observing Interaction: An Introduction to Sequential Analysis*, 2nd edn. Cambridge University Press, Cambridge.

Bales, R.F. (1950) *Interaction Process Analysis: A Method for the Study of Small Groups*. Addison-Wesley Press, Cambridge, MA.

Bales, R.F. (1953) The equilibrium problem in small groups. In: T. Parsons, R.F. Bales, E.A. Snils (eds) *Working Papers in the Theory of Action*. Free Press, New York, pp. 111–163.

Bales, R.F. (1958) Task roles and social roles in problem-solving groups. In: E.E. Maccoby, T.M. Newcomb, E.L. Hartley (eds) *Readings in Social Psychology*, 3rd edn. Holt, New York. pp. 437–447.

Bales, R.F. (1970) *Personality and Interpersonal Behaviour*. Holt, Rinehart & Winston, New York.

Bales, R.F. (1980) *SYMLOG Case Study Kit: With Instructions for a Group Self Study*. The Free Press, New York.

Bales, R.F. & Strodtbeck, F.L. (1951) Phases in group problem-solving. *Journal of Abnormal and Social Psychology*, **XLVI** (46), 485–495.

Bales, R.F., Strodtbeck, F.L., Mills, T.M., Roseborough, M.E. (1951) Channels of communication in small groups. *American Sociological Review*, **16**, 461–468

Bales, R.F, Cohen, S.P., with Williamson, A. (1979) *SYMLOG: A System for the Multiple Level Observation of Groups*. The Free Press, New York.

Banwell, Sir Harold (1964) *The Placing and Management of Contracts for Building and Civil Engineering* Work. HMSO, London.

Barge, J.K. & Keyton, J. (1994). Contextualizing power and social influence in groups. In: L.R. Frey (ed.) *Group Communication in Context: Studies of Natural Groups*. Lawrence Erlbaum Associates, Hove, pp. 85–105.

Barnard, C. (1938) *The Functions of the Executive*. Harvard University Press, Cambridge, MA.

Barrett, J.H. (1970) *Individual Goals and Organisational Objectives: a Study of Integration Mechanisms*. Center for Research on Utilization of Scientific Knowledge. University of Michigan. Ann Arbor, MI.

Bell, L. (2001) Patterns of interaction in multidisciplinary child protection teams in New Jersey. *Child Abuse and Neglect*, **25**, 65–80.

Bemm, D.J, Wallach, M.A., Kogan, N. (1970) Group decision-making under risk of aversive consequences. In: P. Smith (ed.) *Group Processes: Selected Readings*. Middlesex, Penguin, pp. 352–366.

Bentley, T. (1994) Facilitation: providing oppportunities for learning. *Journal of European Industrial Training*, **18** (5), 8–22. MCB University Press.

Blyth, A. & Worthington, J. (2001) *Managing the Brief for Better Design*. Spon Press, London.

Boisot, M.H. (1998) *Knowledge Assets: Securing Competitive Advantage in the Information Economy*. Oxford University Press, Oxford.

Borgatta, E.F. & Bales, R.F. (1953) Task and accumulation of experience as factors in the interaction of small groups. *Sociometry*, **16**, 239–252.

Bormann, E.G. (1996) Symbolic convergence theory and communication in group decision-making. In: R.Y Hirokawa & M.S. Poole (eds) *Communication and Group Decision Making*, 2nd edn. Sage, London, pp. 81–113.

Bowen, P.A. (1993) *A Communication-based Approach to Price Modeling and Price Forecasting in the Design Phase of the Traditional Building Procurement Process in South Africa*. PhD Thesis, University of Port Elizabeth.

Bowen, P.A. (1995) *A Communication-based Analysis of the Theory of Price Planning and Price Control*. RICS research paper, **1**, 2. RICS, London.

Bowen, P.A. & Edwards, P.J. (1996) Interpersonal communication in cost planning during the building design phase. *Construction Management and Economics*, 395–404.

Bowley, M. (1966) *The British Building Industry: Four Studies in Response and Resistance to Change*. Cambridge University Press, Cambridge.

Boyd, D. & Pierce, D. (2001) Implicit knowledge in construction professional practice. *Association of Researchers in Construction Management, 17th Annual Conference*. September 5–7, University of Salford, pp. 37–46.

Broadbent, G. (1973) (1988 revised reprint) *Design in Architecture: Architecture and the Human Sciences*. David Fulton, London.

Brown, I.C. (1963) *Understanding Other Cultures*. Prentice Hall, Englewood Cliffs, NJ.

Brown, R. (2000) *Group Process: Dynamics Within and Between Groups*, 2nd edn. Blackwell Publishers, Oxford.

Brownell, H., Pincus, D., Blum, A., Rehak, A., Winner, E. (1997) The effects of right hemisphere brain damage on patients' use of terms of personal reference. *Brain and Language*, **57**, 60–79.

Brunton, J., Baden Hellard, R., Boobyer, E.H. (1964) *Management Applied to Architectural Practice*, George Godwin for The Builder, London.

Building Industry Communications (1966) *Interdependence and Uncertainty: A Study of the Building Industry*. Tavistock, London.

Burgoon, M., Humsaker, F.G., Dawson, E.J. (1994) *Human Communication*, 3rd edn. Sage, London.

Burke, P.J. (1974) Participation and leadership in small groups. *American Sociological Review*, **39**, December, 832–843.

Calvert, R.E., Bailey, G., Coles, D. (1995) *Introduction to Building Management*, 6th edn. Laxton's, Oxford.

Campbell, A.C. (1968) Selectivity in problem-solving. *American Journal of Psychology*, **81**, 543–550.

Capers, B. & Lipton, C. (1993) Hubble space telescope disaster. *Academy of Management Review*, **7** (3), 23–27.

Cartwright, D. & Zander A. (1962) *Group Dynamics: Research and Theory*. Row Peterson, Evanston, IL.

Cassell, C. & Symon, G. (1994) *Qualitative Methods in Organizational Research*. Sage, London.

Chappell, D. (1996) *Report Writing for Architects and Project Managers*, 3rd edn. Blackwell Science, Oxford.

Clampitt, P.G. and Downs, C.W. (1993) Employee perceptions of the relationship between communication and productivity: a field study. *Journal of Business Communications*, **30**, 5–28.

Clark, R.A. (1993) *Studying Interpersonal Communication: The Research Experience*. Sage, London.

Cline, R.J.W. (1994) Groupthink and the Watergate cover-up: the illusion of unanimity. In: L.R. Frey (ed.) *Group Communication in Context: Studies of Natural Groups*. Lawrence Erlbaum Associates, New Jersey, pp. 199–223.

Collaros, P.A. & Anderson, L.R. (1969) Effect of perceived expertness upon creativity of members of brain-storming groups. *Journal of Applied Psychology*, **53** (2), pt.1, 159–163.

Cook, T.D, Grunder, C.L., Hennigan, K.M., Flay, B.R. (1979) History of the sleeper effect: some logical pitfalls in accepting the null hypothesis. *Psychological Bulletin*, **86**, 662–679.

Dahle, T.L. (1953) *An Objective and Comparative Study of Five Methods of Transmitting Information to Business and Industrial Employees*. PhD thesis, Purdue University.

Dainty, A.R.J. & Moore, D.R. (2000a) Work-group communication problems in design and build project teams: An investigative framework. *Journal of Construction Procurement*, **6** (1), 44–55.

Dainty, A.R.J. & Moore, D.R. (2000b) The performance of integrated D&B project teams in unexpected change event management. *Proceedings of the Association of Researchers in Construction Management, 16th Annual conference.* September 6–8, Glasgow Caledonian University, pp. 281–289.

Daly, J.A., McCroskey, J.C., Ayres, J., Hopf. T., Ayres, D.M. (1997) *Avoiding Communication: Shyness, Reticence and Communication Apprehension*, 2nd edn. Hampton Press, Cresskill, NJ.

Dance, F.E.X. & Larson, C.E. (1972) *Speech Communication: Concepts and Behaviour.* Holt, Rinehart & Winston, New York.

Deal, T.E. & A.A. Kennedy. (1982) *Corporate Cultures: The Rites and Rituals of Corporate Life.* Addison-Wesley, Reading, MA.

De Grada, E., Kruglanski, A.W., Mannetti, L., and Pierro, A. (1999) Motivation cognition and group interaction: need for closure affects the contents and processes of collective negotiations. *Journal of Experimental Social Psychology,* **35**, 346–365.

Dimbleby, R. & Burton, G. (1992) *More than Words: An Introduction to Communication,* 2nd edn. Routledge, London.

Di Salvo, V.S. (1980) Instructional practices: summary of current research identifying communication skills in various organizational contexts. *Communication Education,* **29**, July, 283–290.

Drucker, P.F. (1995) *Managing in a Time of Great Change.* Truman Tulley Books/Dutton, New York.

DuBrin, A.J. (1974) *Fundamentals of Organisational Behaviour: An Applied Perspective.* Pergamon Press, New York.

Duncan, S. Jr. & Fiske, D.W. (1977) *Face-to-face Interaction: Research, Methods and Theory.* Lawrence Erlbaum Associates, New Jersey.

Edland, A. & Svenson, O. (1993) Judgement and decision-making under time pressure: studies and findings. In: O. Svenson & A. J Maule (eds) *Time Pressure and Stress in Human Judgement and Decision Making.* Plenum, New York.

Egan, G. (1973) *Face-to-face: The Small Group Experience and Interpersonal Growth.* Brooks Cole Publishing, Monterey, CA.

Egan, J. (1998) *Rethinking Construction: The Report of the Construction Task Force,* London, DETR. July.

Egan, J. (2002) *Rethinking Construction: Accelerating Change.* Strategic Forum for Construction, London.

Egbu, C. (2000) Knowledge management in construction SMEs: coping with the issues of structure, culture, commitment and motivation. In: A. Akintoye (ed.) *The Proceedings of the 16th Annual Conference of the Association of Researchers in Construction Management.* September 6–8, Glasgow Caledonian University, pp. 83–92.

Egbu, C., Botterill, K., Bates, M. (2001) The influence of knowledge management and intellectual capital on organizational innovations. In: A. Akintoye (ed.) *The Proceedings of the 17th Annual Conference of the Association of Researchers in Construction Management.* September 5–7, Universtiy of Salford, pp. 547–556.

Ellis, D.G. & Fisher, B.A. (1994) *Small Group Decision Making: Communication and the Group Process,* 4th edn. McGraw-Hill, New York.

Emmerson, H. (1962) *Survey of Problems Before the Construction Industries: A Report Prepared for the Minister of Works.* HMSO, London.

Emmitt, S. (1994) 'Keeper of the gate'. *International Journal of Architectural Management, Practice and Research,* **8**, 23–26.

Emmitt, S. (1997) *The Diffusion of Innovations in the Building Industry,* PhD thesis, University of Manchester.

Emmitt, S. (1999) *Architectural Management in Practice: A Competitive Approach.* Longman, Harlow.

Emmitt, S. (2002) *Architectural Technology.* Blackwell Science, Oxford.

Emmitt, S. & Gorse, C.A. (1996) De-construction procurement routes opening up for detail. In: S. Emmitt (ed.) *Detail Design in Architecture.* BRC, Northampton.

Emmitt, S. & Wyatt, D.P. (2000) Worlds within worlds: networking towards ecological building, *Sustainable Product Information,* CIB 102, Helsinki, Finland.

Emmitt, S. & Yeomans, D.T. (2001) *Specifying Buildings: A Design Management Perspective.* Butterworth-Heinemann, Oxford.

Farmer, S.M. & Roth, J. (1998) Conflict-handling behavior in work groups. *Small Group Research*, **29** (6), 669–698.

Faulkner, A.C. & Day, A.K. (1986) Images of status and performance in building team occupations. *Construction Management and Economics*, **4**, 245–260.

Feldberg, M. (1975) *Organization Behaviour: Text and Cases*. Juta and Company, Cape Town.

Fielding, N.G. & Fielding, J.L. (1986) *Linking Data*. Sage Publications, London.

Fiske, J. (1990) *Introduction to Communication Studies*, 2nd edn. Routledge, London.

Fledman, D.C. (1984) The development and enforcement of group norms. *Academy of Management Review*, **9**, 47–53.

Frey, L.R. (1999) *The Handbook of Group Communication Theory and Research*. Sage, London.

Frost, P.J. (1987) Power, politics and influence. In: F.M. Jablin, L.L. Putnam, K.H. Roberts, L.W. Porter (eds) *Handbook of Organizational Communication*. Sage, Beverly Hills, CA, pp. 503–548.

Fryer, B., Egbu, C., Ellis, R., Gorse, C. (2003) *The Practice of Construction Management*, 4th edn. Blackwell Publishing, Oxford.

Gameson, R.N. (1992) *An Investigation into the Interaction between Potential Building Clients and Construction Professionals*. PhD thesis, University of Reading.

Gardiner, P.D. & Simmons, J.E.L. (1992) Analysis of conflict and change in construction projects. *Construction Management and Economics*, **10**, 459–478.

Gibb, J.R. (1961) Defensive communication. *Journal of Communication*, **11**, 141–148.

Giles, H. (1986) General preface. In: W.B. Gudykunst (ed.) *Intergroup Communication: The Social Psychology of Language 5*. Edward Arnold Publishers, Baltimore.

Glass, A.L. & Holyoak, K.J. (1986) *Cognition*, 2nd edn. McGraw Hill, New York.

Goleman, D. (1996) *Emotional Intelligence*. Bloomsbury, London.

Gorse, C.A. (2002) *Effective Interpersonal Communication and Group Interaction During Construction Management and Design Team Meetings*. PhD thesis, University of Leicester.

Gorse, C.A. & Emmitt, S. (1998) Information exchange between the architect and the contractor. In: S. Emmitt (ed.) *The Product Champions: Proceedings of the 2nd International Conference on Detail Design in Architecture*. November, pp. 182–199.

Gorse, C.A., Emmitt, S., Lowis, M. (1999) Problem solving and appropriate communication medium. In: W. Hughes, *Association of Researchers in Construction Management, 15th Annual Conference*. September 15–17, Liverpool, John Moores University, pp. 511–518.

Gorse, C.A., Emmitt, S. Lowis, M., Howarth, A. (2000a) A critical examination of methodologies for studying human communication in the construction industry. *Association of Researchers in Construction Management, 16th Annual Conference*. September 6–8, Glasgow Caledonian University, pp. 31–39.

Gorse, C.A., Emmitt, S., Lowis, M., Howarth, A. (2000b). Interaction analysis during management and design team meetings. *Association of Researchers in Construction Management, 16th Annual Conference*. September 6–8, Glasgow Caledonian University, pp. 763–771 (2 vols).

Gorse, C.A., Emmitt, S. Lowis, M., Howarth, A. (2000c) A methodology for research of construction communication. In S. Emmitt (ed.) *Detail Design in Architecture 3*, September 12–13, Brighton University, pp. 41–50.

Gorse, C.A., Emmitt, S. Lowis, M., Howarth, A. (2000d) Models of group and interpersonal interaction during management and design team meetings. In S. Emmitt (ed.) *Detail Design in Architecture 3*, September 12–13, Brighton University, pp. 50–60.

Gorse, C.A., Emmitt, S., Lowis, M., Howarth, A. (2001) *Project Performance and Management and Design Team Communication*. In: A. Akintoye (ed.) *Proceedings of the Association of Researchers in Construction Management, 17th Annual Conference*, September 5–7, University of Salford.

Gorse, C.A., Emmitt, S., Lowis, M., Howarth, A. (2002) Interaction characteristics of successful contractor's representatives. In D. Greenwood (ed.) *Proceedings of the Association Researchers in Construction Management, 18th Annual Conference*, September 2–4, Northumbria University, pp. 187–188.

Gower, E. (1954) *The Complete Plain Words* (revised by S. Greenbaum & J. Whitcut, 1986), Penguin Books, London.

Green, A. Thorpe, T., Austin, S. (2002) As likely as not it could happen: linguistic interpretations of risk. In: D. Greenwood (ed.) *Proceedings of the 18th Annual Conference of the Association of Researchers in Construction Management*. Northumbria University, pp. 637–646.

Hackman, J.R. (ed.) (1990) *Groups that Work (and Those that Don't): Creating Conditions for Effective Teamwork.* Jossey-Bass, San Francisco.

Hackman, J.R (1992) Group influences on individuals in organizations. In: M.D. Dunnette & L.M. Hough (eds) *Handbook of Industrial and Organizational Psychology*, 2nd edn. Consulting Psychologists' Press, Palo Alto, CA.

Hackman, J.R. & Vidmar, N. (1970) Effects of size and task type on group and member reactions. *Sociometry*, **33**, 1 (March), 37–54.

Hall, M. (2001) 'Root' cause analysis: a tool for closer supply chain integration in construction. *Association of Researchers in Construction Management, 17th Annual Conference*, September 5–7, University of Salford, pp. 929–938.

Hancock, R.D. & Sorrentino, R.M. (1980) The effects of expected future interactions and prior group support on the conformity process. *Journal of Experimental Social Psychology*, **16** (3), 261–269.

Handy, C.B. (1981) *Understanding Organisations.* Penguin, London.

Hare, A.P. (1976) *Handbook of Small Group Research*, 2nd edn. The Free Press, New York.

Hargie, O.D.W, Dickson, D., Tourish, D. (1999) *Communication in Management.* Gower, Hampshire.

Harper, D.R. (1978) *Building: The Process and the Product.* The Construction Press, Lancaster.

Hartley, P. (1997) *Group Communication.* Routledge, London.

Hastings, I. (1998) The virtual project team. *Project Manager Today*, July, 26–29.

Heinicke, C. & Bales, R.F. (1953) Developmental trends in the structure of small groups, *Sociometry*, **16**, 7–38.

Hertzberg, F. (1966) *Work and the Nature of Man.* Collins, New York.

Higgin, G. & Jessop, N. (1965) *Communications in the Building Industry: The Report of a Pilot Study.* Tavistock, London.

Hill, C.J. (1995) Communication on construction sites, *Association of Researchers in Construction Management, 11th Annual Conference*, September 18–20, University of York, pp. 232–240.

Hirokawa, R.Y. & Poole, M.S. (1996) *Communication and Group Decision Making.* Sage, London.

Hirokawa, R.Y. & Salazar, A.J. (1999) Task-group communication and decision-making performance. In: L.R. Frey (ed.) *The Handbook of Group Communication Theory and Research.* Sage, London, pp. 167–191.

Hirokawa, R.Y. & Scheerhorn, D.R. (1986) Communication in faulty group decision-making. In: R.Y. Hirokawa & M.S. Poole (eds) *Communication and Group Decision-making.* Sage, Beverly Hills, pp. 63–80.

Hollingshead, A.B. (1996) The rank order effect in group decision-making, *Organizational Behaviour and Human Decision Processes*, **68**, 3 (December), 181–193.

Hollingshead, A.B. (1998) Communication, learning and retrieval in transactive memory systems. *Journal of Experimental Social Psychology*, **34**, 423–442.

Hosking, D. & Haslam, P. (1997) Managing to relate: organizing as a social process. *Career Development International*, **2** (2), 85–89.

Hughes, W.P. (1989) *Organisational Analysis of Building Projects.* PhD thesis, Liverpool Polytechnic.

Hughes, W.P. (1994) The PhD in construction management. *Association of Researchers in Construction Management, 10th Annual Conference*, September 14–16, Loughborough University of Technology, pp. 76–87.

Hugill, D. (1998) Illuminating a psychological theory (in a construction management context). *Association of Researchers in Construction Management, 14th Annual Conference.* September 9–11, University of Reading, pp. 22–30.

Hugill, D. (1999) Negotiating access: presenting a credible project. In: W. Hughes (ed.), *Proceedings of the 15th Annual Conference of the Association of Researchers in Construction Management*, September 15–17, Liverpool, John Moores University, pp. 53–63.

Hugill, D. (2000) Management as an accomplishment of project team meetings in construction. In: A. Akintoye (ed.) *Proceedings of the 16th, Annual Conference of the Association of Researchers in Construction Management*, September 6–8, Glasgow Caledonian University, pp. 755–762.

Hugill, D. (2001) *An Examination of Project Management Team Meetings in Railway Construction.* PhD thesis, University of Manchester.

Huseman, R.C. (1977) Interpersonal conflict in the modern organisation. In: R.C. Huseman,

C.M. Logue, D.L. Freshley, *Readings in Interpersonal and Organisational Communication*, 3rd edn. Allyn & Bacon, London.

Huseman, R.C., Logue. C.M., Freshley, D.L. (1977) *Readings in Interpersonal and Organisational Communication*, 3rd edn. Allyn & Bacon, London.

Janis, I.L. (1982) *Groupthink: Psychological Studies of Policy Decisions and Fiascos*, 2nd edn. Houghton Mifflin, Boston, MA.

Jobber, D. (1991) Choosing a survey method in management research. In: N.C. Smith & P. Dainty (eds) *The Management Research Handbook*. Routledge, London, pp. 174–180.

Kaderlan, N. (1991) *Designing Your Practice: A Principal's Guide to Creating and Managing a Design Practice*. McGraw-Hill, New York.

Katz, D. & Kahn, R.L. (1967) *The Social Psychology of Organizations*. John Wiley & Sons, New York.

Ketrow, S.M. (1999) Nonverbal aspects of group communication. In: L.R. Frey (ed.) *The Handbook of Group Communication Theory and Research*. Sage, London, pp. 251–287.

Keyton, J. (1999) Relational communication in groups. In: L.R. Frey (ed.) *The Handbook of Group Communication Theory and Research*. Sage, London, pp. 192–221.

Keyton, J. (2000) Introduction: the relational side of groups. *Small Group Research*, **31** (4), 387–394.

Kolb, D. (1992) *Hidden Conflict in Organisations*. Sage Publications, London.

Kotter, J.P. and Schlesinger, L.A. (1979) Choosing strategy for change. *Harvard Business Review*, March/April, 106–114.

Kreps, G.L. (1984) *Using the Case Study Method in Organizational Communication Classes: Developing Students' Insight, Knowledge and Creativity*. International Communication Association, San Francisco.

Kreps, G.L. (1989) *Organizational Communication: Theory and Practice*, 2nd edn. Longman, New York.

Larson, C.W. & LaFasto, F.M.J. (1989) *Teamwork: What Must Go Right/What Can Go Wrong*. Sage, Beverly Hills.

Latham, M. (1993) *Trust and Money: Interim Report of the Joint Government Industry Review of Procurement and Contractual Arrangements in the United Kingdom Construction Industry*. HMSO, London.

Latham, M. (1994) *Constructing the Team*. Final Report, HMSO, London.

Lavers, A.P. (1992) Communication and clarification between designer and client: good practice and legal obligation. In: M.P. Nicholson, *Architectural Management*. E & FN Spon, London.

LeDoux, J. (1998) *The Emotional Brain*. Phoenix, New York.

Lee, F. (1997) When the going gets tough, do the tough ask for help? Help-seeking and power motivation in organizations. *Organizational Behaviour and Human Decision Processes*, **72** (3) (December), 336–363.

Lewin, K. (1947) Frontiers in group dynamics II: channels of group life, social planning and action research. *Human Relations*, **1**, 143–153.

Lewin, K. (1951) *Field Theory in Social Science*. Harper & Row, New York.

Lieberman, M., Lakin, M., Whitaker, D. (1969) Problems and potential of psychoanalytic and group theories for group psychotherapy. *International Journal of Group Psychotherapy*, **19**, 131–141.

Littlepage, G.E. & Silbiger, H. (1992) Recognition of expertise in decision-making groups: effects of group size and participation patterns. *Small Group Research*, **22**, 344–355.

Lonetto, R. & Williams, D. (1974) Personality, behavioural and output variables in a small group task situation: an examination of consensual leader and non-leader differences. *Canadian Journal of Behavioural Science*, **6**, 58–74.

Loosemore, M. (1992) Managing the construction process through a framework of decisions. In M.P. Nicholson, *Architectural Management*. Spon, London, pp. 90–103.

Loosemore, M. (1994) Problem behaviour. *Construction Management and Economics*, **12**, 511–520.

Loosemore, M. (1996) *Crisis Management in Building Projects: A Longitudinal Investigation of Communication Behaviour and Patterns within a Grounded Framework*. PhD thesis, University of Reading.

Loosemore, M. (1998) The influence of communication structure upon crisis management efficiency. *Construction Management and Economics*, **16**, 661–671.

Loosemore, M. (1999) Responsibility, power and construction conflict, *Construction Management and Economics*, **17**, 699–709.

Loosemore, M. & Chin Chin Tan (2000) Occupational stereotypes in the construction industry. *Construction Management and Economics*, **18**, 559–566.

Loosemore, M., Nguyen, B.T., Denis, N. (2000) An investigation into the merits of encouraging conflict in the construction industry. *Construction Management and Economics*, **18**, 447–456.

Luft, J. (1984) *Group Process: An Introduction to Group Dynamics*. Mayfield, Palo Alto, CA.

Mabry, E.A. (1999) The systems metaphor in group communication. In: L.R. Frey (ed.) *The Handbook of Group Communication Theory and Research*. Sage, London, pp. 71–91.

Mackinder, M. & Marvin, H. (1982) *Design Decision Making in Architectural Practice*, Research Paper 19. University of York Institute of Advanced Architectural Studies, York.

Maister, D.H. (1993) *Managing the Professional Service Firm*. The Free Press, New York.

Maslow, A.H. (1954) *Motivation and Personality*. Harper & Row, New York.

Masnikosa, V.P. (1999) On some obstacles in communication and transfer of knowledge. *Kybernetes*, **28** (5), 575–584.

McCann, D. (1993) *How to Influence Others at Work: Psychoverbal Communication for Managers*, 2nd edn. Butterworth-Heinemann, London.

McCroskey, J.C. (1977a) Oral communication apprehension: a summary of recent theory and research. *Human Communication Research*, **4**, 78–96.

McCroskey, J.C (1997b) Willingness to communicate, communication apprehension, and self-perceived communication competence: conceptualizations and perspectives. In: J.A. Daly, J.C. McCroskey, J. Ayres, T. Hopf, D.M. Ayres (eds) *Avoiding Communication: Shyness, Reticence and Communication Apprehension*. Hampton Press, New Jersey, pp. 75–108.

McCroskey, J.C. & Richmond, V.P. (1990). Willingness to communicate: a cognitive view. *Journal of Social Behavior and Personality*, **5**, 19–37.

McGrath, J.E. (1984) *Groups: Interaction and Performance*. Prentice Hall, Englewood Cliffs, NJ.

Mellinger, G.D. (1956) Interpersonal trust as a factor in communication. *Journal of Abnormal and Social Psychology*, **52**, 22–30.

Meyers, R.A. & Brashers, D.E. (1999) Influence processes in group interaction. In: L.R. Frey (ed.) *The Handbook of Group Communication Theory and Research*. Sage, London, pp. 288–312.

Middleton, D. (1996) Talking work: argument, common knowledge and improvisation in Teamwork. In: Y. Engestrom & D. Middleton (eds) *Cognition and Communication at Work*. Cambridge University Press, Cambridge.

Millet, S.J., Dainty, A.R., Briscoe, G.H. (2001) Supply chain management: is there a relationship between the procurement of clients' core services and its construction procurement? *International Postgraduate Research in the Built and Human Environment*. University of Salford, pp. 14–23.

Mintzberg, H. (1973) Strategy making in three models. *California Management Review*, Winter, 44–53.

Mior, Azam, M.A., Ross, A.D., Fortune, C.J., Jagger, D. (1998) An information strategy to support effective construction design decision making. *Association of Researchers in Construction Management, 14th Annual Conference*. September 9–11, University of Reading, pp. 248–257.

Moore, D.R. & Dainty, A.R.J (2002) Criticizing the 'techniques of communication approach': a response. In: D. Greenwood (ed.) *Proceedings of the 18th Annual Conference of the Association of Researchers in Construction Management*. Northumbria University, pp. 237–236.

Morrison, E.W. (1993) Newcomer information seeking: exploring types, models, sources and outcomes. *Academy of Management Journal*, **36** (3), 556–589.

Moser, C.A. and Kalton, G. (1971) *Survey Methods in Social Investigation*. Heinemann, London.

Mullen, B., Salas, E., Driskell, J.E. (1989) Salience, motivation and artefact as contributions to the relation between participation rate and leadership. *Journal of Experimental Social Psychology*, **25**, 545–559.

Murray, M.A. (1976) Education for public administrators. *Public Personnel Management*, **5**, New York.

Needham, M.J. (1998) Arbitration for the Construction Industry. Annual guest lecture series, at Leeds Metropolitan University, School of the Built Environment, unpublished paper, 25 March 1998.

Nicholson, M.P. (ed.) (1992) *Architectural Management*. E. & F.N. Spon, London.

Nicholson, P. (1997) Design and build is not enough. *Construction Manager*, **3** (8), 55.

Ordonez, L. & Benson, L. (1997) Decisions under time pressure: how time constraints affect risky decision-making. *Organizational Behaviour and Human Decision Processes*, **71** (2), 121–140.

Ottaway, R.N. (1982) Defining the change agent. In: B. Evans, J.A. Powell, R. Talbot (eds.) *Changing Design*. John Wiley & Sons, Chichester.

Patchen, M. (1993) Reciprocity of coercion and co-operation. In: R.B. Felson & J.T. Tedeschi (eds) *Aggression and Violence: Social Interactionist Perspectives*. American Psychological Association, Washington, pp. 119–144.

Paterson, J. (1977) *Information Methods: For Design and Construction*. John Wiley, London.

Pavitt, C. (1999) Theorizing about the group communication–leadership relationship. Input–process–output and functional models. In: L.R. Frey (ed.) *The Handbook of Group Communication Theory and Research*. Spon, London, pp. 313–334.

Philipsen, G. & Albrecht, T.L. (1997) *Developing Communication Theories*. State University Press, New York.

Phillips Report (1950) *Report of a Working Party to the Minister of Works*, HMSO, London.

Pietroforte, R. (1992) *Communication and Information in the Building Delivery Process*. PhD thesis, Massachusetts Institute of Technology, Cambridge, MA.

Pietroforte, R. (1997) Communication and governance in the building process. *Construction Management and Economics*, **15**, 71–82.

Pondy, L. (1967) Organisational conflict, concepts and models. *Administrative Quarterly*, 12 (September), 299–306.

Poole, M.S. (1981) Decision development in small groups, I: a comparison of two models. *Communication Monographs*, **48**, 1–24.

Poole, M.S. (1983a) Decision development in small groups, II: a study of multiple sequences in decision making. *Communication Monographs*, **50**, 206–232.

Poole, M.S. (1983b) Decision development in small groups, III: a multiple sequence model of decision development. *Communication Monographs*, **50**, 321–341.

Poole, M.S. (1999) Group communication theory. In: L.R. Frey (ed.) *The Handbook of Group Communication Theory and Research*. Sage, London, pp. 37–70.

Poole, M.S. & Baldwin, C.L. (1996) Developmental processes in group decision-making. In: R.Y. Hirokawa & M.S. Poole (eds) *Communication and Group Decision Making*, 2nd edn. Sage, Thousand Oaks, California, pp. 215–241.

Poole, M.S. & Hirokawa, R.Y. (1996) Communication and group decision-making. In: R.Y. Hirokawa and M.S. Poole (eds) *Communication and Group Decision Making*. Sage, London, pp. 3–18.

Poole, M.S., Keyton, J. & Frey, L.R. (1999) Group communication methodology: issues and considerations. In: L.R. Frey (ed.) *The Handbook of Group Communication Theory and Research*. Sage, London, pp. 92–112.

Potter, J. & Wetherell, M. (1987) *Discourse and Social Psychology: Beyond Attitudes and Behaviour*. London, Sage.

Potter, N. (1989) *What is a Designer?: Things, Places, Messages*, 3rd edn. Hyphen Press, London.

Preece, C. & Stocking, S. (1999) Safety communications management in construction contracting. *Association of Researchers in Construction Management, Proceedings of the 15th Annual Conference*. September 15–17, Liverpool, John Moores University, pp. 529–539.

Price, S. (1996) *Communication Studies*. Longman, Harlow.

Pruitt, D.G, Mikolic, J.M, Peirce, R.S., Keating, M. (1993) Aggression as a struggle tactic in social conflict. In: R.B. Felson & J.T. Tedeschi (eds) *Aggression and Violence: Social Interactionist Perspectives*. Harvard, American Psychological Association, pp. 99–118.

Rahim, M.A. (1983) A measure of styles of handling interpersonal conflict. *Academy of Management Journal*, **26** (2), 368–376.

Rasberry, R.W. & Lindsay, L.L. (1989) *Effective Managerial Communication*, 2nd edn. Wadsworth Publishing Co, California.

Ribeiro, F.L. & Lopes, J. (2001) Construction supply chain integration over the Internet and web technology. In: A. Akintoye (ed.) *Association of Researchers in Construction Management, 17th Annual Conference.* September 5–7, University of Salford, pp. 241–251.

Richard, D. & Kroeger, L. (1989) Messages clear and effective. *Journal of Management in Engineering,* **5** (2), 186–191.

Rim, Y. (1964) Social attitudes and risk-taking. *Human Relations,* **17**, 259–265.

Rim, Y. (1965) Leadership attitudes and decisions involving risk. *A Journal of Applied Research,* **18**, 423–430.

Rim, Y. (1966) Machiavellianism and decisions involving risk. *British Journal of Social and Clinical Psychology,* **5**, 30–36.

Robson, C. (1993) *Real World Research: A Resource for Social Scientists and Practitioners-Researchers.* Blackwell, Oxford.

Rogers, E.M. (1986) *Communication Technology: The New Media in Society.* The Free Press, New York.

Rogers, E.M. (1995) *Diffusion of Innovations,* 4th edn. The Free Press, New York.

Rogers, E.M. & Kincaid, D.L. (1981) *Communication Networks: Toward a New Paradigm for Research.* The Free Press, New York.

Roodman, H. & Roodman, Z. (1973) *Management by Communication.* Methuen Publications, Toronto.

Royal Institute of British Architects (1962) *The Architect and His Office.* RIBA, London.

Scheidel, T.M. & Crowell, L. (1964) Idea development in small discussion groups. *Quarterly Journal of Speech,* **50**, 140–145.

Scheidel, T.M. & Crowell, L. (1966) Feedback in group communication. *Quarterly Journal of Speech,* **52**, 273–278.

Schultz, B.C. (1999) Improving group communication performance, an overview of diagnosis and intervention. In: L.R. Frey (ed.) *The Handbook of Group Communication Theory and Research.* Sage, London, pp. 371–394.

Scott, C.R. (1999) Communication technology and group communication. In: L.R. Frey (ed.) *The Handbook of Group Communication Theory and Research.* Sage, London, pp. 432–474.

Seymour, D. & Hill, C. (1993) Implications of research perspective for management policy. *Association of Researchers in Construction Management, 9th Annual Conference.* September 14–16, pp. 151–123.

Shannon, C.E. & Weaver, W. (1949) *The Mathematical Theory of Communication.* University of Illinois, Urbana.

Shapiro, D. & Leiderman, P.H. (1967) Arousal correlates of task role and group setting, *Journal of Personality and Social Psychology,* **5** (1), 103–107.

Shaw, B.M. (1957) Decision process in communication nets. *Journal of Abnormal Social Psychology,* **54**, 323–330.

Shaw, M. (1981) *Group Dynamics: The Psychology of Small Group Behaviour,* 3rd edn. McGraw-Hill, New York.

Shepherd, C.R. (1964) *Small Groups: Some Sociological Perspectives.* Chandler Publishing, New York.

Shoemaker, P.J. (1991) *Gatekeeping.* Sage Publications, Newbury Park, CA.

Simmel, G. (1946) *The Web of Group-Affiliations* (Trans. R. Bendix). Free Press, New York.

Simon, E.D. (1944) *The Placing and Management of Building Contracts.* HMSO, London.

Simon, E.D. (1945) *Rebuilding Britain: A Twenty-Year Plan.* Victor Gollancz New Left Books, London.

Simon, E.D. (1948) *The Distribution of Building Materials and Components.* HMSO, London.

Slater, R.E. (1958) Thinking ahead. *Harvard Business Review,* **36** (2), 27–39.

Smith, J. & Wyatt, R. (1998) Criteria for strategic decision making at the pre-briefing stage. *Association of Researchers in Construction Management, 14th Annual Conference,* September 9–11, University of Reading, pp. 300–309.

Smith, R.L., Richetto, G.M., & Zima, J.P. (1977) Organizational behaviour: an approach to human communication. In: R.C. Huseman, C.M. Logue, D.L. Freshley (eds) *Readings in Interpersonal and Organisational Communication,* 3rd edn. Allyn & Bacon, London.

Spence, W.R. (1994) *Innovation: The Communication of Change in Ideas, Practices and Products.* Chapman & Hall, London.

Sperber, D. & Wilson, D. (1986) *Relevance Communication and Cognition.* Blackwell, Oxford.

Stewart, R. (1967) *Managers and Their Jobs*. Macmillan, London.

Stone, P., Dunphy, D.C., Smith, M.S. & Ogilvie, D.M. (1966) *The General Enquire: A Computer Approach to Content Analysis*. MIT Press, Cambridge, MA.

Stretton, A. (1981) Construction communications and individual perceptions. *Chartered Builder*, Summer, 51–53.

Stroop, J.R. (1932) Is the judgement of the group better than that of the average member of the group? *Journal of Experimental Psychology*, **15**, 550–562.

Symon, G. (1998) Qualitative research diaries. In: G. Symon & C. Cassell (eds) *Organisational Research: A Practical Guide*. Sage, London.

Tarde, G. (1903) *The Laws of Imitation* (Trans. E. Clews Parsons) (cited in Rogers 1995). Holt, New York.

Thelen, H.A. (1949) Group dynamics in instruction: principle of least group size. *School Review*, **57**, 139–148.

Thomas, K.W. (1976) Conflict and conflict management. In: M.D. Dunnette (ed.) *Handbook of Industrial and Organizational Psychology*. Rand McNally, Chicago, pp. 889–935.

Thomas, K.W. & Kilmann, R.H. (1975) The social desirability variable in organizational research: an alternative explanation for reported findings. *Academy of Management Journal*, **18** (4), 141–153.

Thomas, S.R., Tucker, R.L., Kelly, W.R. (1998) Critical communication variables, *Journal of Construction Engineering and Management*. **124** (1), 58–66.

Trenholm, S. & Jensen, A. (1995) *Interpersonal Communication*, 3rd edn. Wadsworth Publishing, London.

Tubbs, S.L. & Moss, S. (1981) *Interpersonal Communication*, 2nd edn. Random House, New York.

Tuckman, B.W. (1965) Developmental sequences in small groups. *Psychological Bulletin*, **63**, 384–399.

Wadleigh, P.M. (1997) Contextualizing communication avoidance research. In: J. Daly, J.C. McCroskey, T. Hopf, D. Ayres (eds) *Avoiding Communication*, 2nd edn. Hampton Press, Cresskill, New Jersey, pp. 3–20.

Walker, A. (2002) *Project Management in Construction*, 4th edn. Blackwell Science, Oxford.

Wallace, W.A. (1987) *The Influence of Design Team Communication Content upon the Architectural Decision Making Process in the Pre-contract Design Stages*. PhD thesis, Heriot-Watt University, Edinburgh.

Watson, P. (2000) Managing people: the key activity for site managers. *Construction Manager*, **5** (10), 48–49.

Westley, B.H. & MacLean, Jr., M.S. (1957) A conceptual model for communications research. *Journalism Quarterly*, **34**, 31–38.

White, D.M. (1950) The 'gate keeper': a case study in the selection of news. *Journalism Quarterly*, **27**, 383–390.

Wild, A. (2001) Construction projects as teams or situations: criticizing the techniques of communication approach. In: A. Akintoye (ed.) *Proceedings of the 17th Annual Conference of the Association of Researchers in Construction Management*, September 5–7, University of Salford, pp. 495–504.

Wild, A. (2002a) 'False coherence' and 'cohesive fragmentation' in construction's post-war corporatism 1945–70. In: D. Greenwood (ed.) *Proceedings of the 18th Annual Conference of the Association of Researchers in Construction Management*, Northumbria University, pp. 729–738.

Wild, A. (2002b) The management college that never was. In: D. Greenwood (ed.) *Proceedings of the 18th Annual Conference of the Association of Researchers in Construction Management*, Northumbria University, 739–748.

Wilmot, W.W. (1980) *Dyadic Communication*, 2nd edn. Addison-Wesley, Reading, MA.

Winch, G. & Schneider, E. (1993) Managing the knowledge-based organisation: the case of architectural practice. *Journal of Management Studies*, **30** (6), 923–937.

Woodcock, B.E. (1979) Characteristics oral and written business communication problems of selected managerial trainees. *Journal of Business Communication*, **16**, 43–49.

Wyatt, N. (1993) Organizing and relating: feminist critique of small group communication. In: S.P. Bowen and N. Wyatt (eds) *Transforming Visions: Feminist Critiques in Communication Studies*. Hampton Press, Cresskill, New Jersey, pp. 21–86.

Wyatt, D.P. & Emmitt, S. (1997) The product's information network impediment. *The International Journal of Architectural Management, Practice and Research*, **13**, 30–38.

Yoshida, R.K., Fentond, K., Maxwell, J. (1978) Group decision making in the planning team process: myth or reality? *Journal of School Psychology*, **16**, 237–244.

Ysseldyke, J.E., Algizzine, B., Mitchell, J. (1982) Special education team decision making: an analysis of current practice. *Personnel and Guidance Journal*, **60** (5), 308–310.

Zahrly, J. & Tosi, H. (1989) The differential effects of organizational induction process on early work role adjustment. *Journal of Organizational Behavior*, **10**, 59–74.

Index